SpringerBriefs in Sociology

More information about this series at http://www.springer.com/series/10410

Tina Sikka

Climate Technology, Gender, and Justice

The Standpoint of the Vulnerable

 Springer

Tina Sikka
Media and Cultural Studies
Newcastle University
Newcastle upon Tyne, UK

ISSN 2212-6368   ISSN 2212-6376   (electronic)
SpringerBriefs in Sociology
ISBN 978-3-030-01146-8   ISBN 978-3-030-01147-5   (eBook)
https://doi.org/10.1007/978-3-030-01147-5

Library of Congress Control Number: 2018957837

This Springer imprint is published by the registered company Springer Nature Switzerland AG
The registered company address is: Gewerbestrasse 11, 6330 Cham, Switzerland

Contents

Chapter 1
Introduction

Abstract This introduction provides preliminary remarks and clarifications on structure and definitions used throughout the text, followed by a few concise explanations of the theoretical frameworks and a robust justification of methodology and choice of literature. A basic definition of geoengineering is provided as well as a discussion of the feminist frameworks deployed throughout the book including feminist contextual empiricism, feminist standpoint theory, and technofeminism.

Keywords Geoengineering · Feminist empiricism · Feminist standpoint theory · Technofeminism · Sociology of science studies

The complicated relationship between gender, natural or physical science, technology and climate change has become a subject of a intense discussion in a variety of settings from academia and governments to civil society and think tanks. Yet most of these bodies tend to focus on one or two subjects often at the expense of others. On the topic of climate change, as a sociological phenomenon, a trans or interdisciplinary approach is needed in order to map complex connections between diverse areas. Because it is difficult to garner enough expertise in order to, for example, answer questions about how the generation of scientific knowledge about climate change affects gender norms, or what role gender plays in the construction of technological artifacts that might revolutionize how we generate power, it is important to be clear about what questions require answering, what gaps in knowledge remain, how this can be addressed, and whose interests does maintaining the status quo serve.

This book aims to provide a set of unique perspectives on the interconnections between the traditionally "silo-ized" categories of gender, climate science, and climate technologies using climate geoengineering as a case study. My objective is to critically assess climate geoengineering science, modeling, and symbolism/discourse, by using contemporary feminist approaches to science and technology studies. In doing so, I seek to expose a unique set of challenges posed by climate change and technological innovation by drawing on feminist contextual empiricism (FCE) primarily but also feminist standpoint theory (FST) and technofeminism. Central to this approach is the thesis that the feminist tenets of diversity, pluralism, situated knowledge, values, community, and sociality pose significant challenges to aggressive solutions to climate change including that of climate geoengineering.

In this introduction, I begin with some preliminary remarks and clarifications on structure and definitions used throughout the text, followed by a few concise explanations of theoretical frameworks and a robust justification of methodology and choice of literature. A basic definition of geoengineering is provided as well as a discussion of the feminist frameworks deployed throughout the book. I maintain that feminist contextual empiricism, as articulated by Helen Longino, provides the most persuasive critique of geoengineering by using feminist science studies. Feminist standpoint theory and technofeminism deliver further insights that make up for a handful of shortcomings of the feminist empiricist approach.

Methodology/Theoretical Approach

Methodologically, the approach I take is, first and foremost, multiperspectival with respect to the contributions of feminist perspectives on science and technology articulated above. While these frameworks have considerable differences in how they take up the relationship between science, technology and gender, harnessing contributions from all three, while focusing predominantly on FCE, provides a theoretically rich analytic framework from which significant insights can be assembled. This study is also disciplinarily pluralist since, in that in addition to pulling from contributions of the above noted feminist perspectives, also incorporated are knowledge and ideas from the natural sciences, computer science, political science, international law, social constructivism and public policy. Yet, rather than resulting in a discordant cacophony of divergent perspectives, this path has been both theoretically fruitful and analytically complementary since,

> ...a thorough-going disciplinary pluralism [...] suggests that sometimes the perspectives don't fit nicely together on the same plane: they overlap or conflict or cannot both be held at the same time, and yet you need both of them (Kellert 2006: 225).

Climate Change and Geoengineering

Over the past few years, a surfeit of scientific studies, research proposals, media stories and government reports have coalesced around a solution to climate change that would employ the latest and most cutting edge technologies. These geoengineering or climate engineering proposals have garnered a particularly high level of attention not only because of increased levels of media coverage, research investment, and government interest, but also in light of more mainstream mitigation strategies, such as emission cuts and renewable energy technologies, being seen as falling short of the 1.5–2 °C warming limit. These technical solutions are constituted by a wide variety of schemes designed to mitigate climate change through direct interventions aimed at either removing carbon dioxide from the atmosphere (and sequestering it), or reflecting solar radiation from the earth itself (thereby removing the ability of heat absorbing radiation to warm our climate).

The case for increased research into and use of renewable energy (wind, solar, geothermal etc.) as a viable, scientifically sound, and less risky path towards addressing climate change, rather than geoenginering, is well established (Panwar et al. 2011; Twidell and Weir 2015; Luderer et al. 2014). Countries around the world have made significant strides in adopting renewables as their price has fallen and investment has risen. Global wind power generation in 2015 hit 7% of total global power generation capacity, natural gas is now the second largest power generation source at 22%, and solar now produces 1% of electricity use globally (World Energy Council 2016). According Ren21,

> Renewable power generating capacity saw its largest annual increase ever in 2016, with an estimated 161 gigawatts (GW) of capacity added. Total global capacity was up nearly 9% compared to 2015, to almost 2,017 GW at year's end. The world continued to add more renewable power capacity annually than it added (net) capacity from all fossil fuels combined. In 2016, renewables accounted for an estimated nearly 62% of net additions to global power generating capacity. Solar PV saw record additions and, for the first time, it accounted for more additional capacity, net of decommissioning, than did any other power generating technology. Solar PV represented about 47% of newly installed renewable power capacity in 2016, and wind and hydropower accounted for most of the remainder, contributing 34% and 15 .5%, respectively (Ren21 2017, 20).

Yet it is also the case that without a significant acceleration of its use, it is coupled with drastic changes in the patterns of production and consumption, there is a very real danger that the 2 °C target will be exceeded this century. This is where geoengineering as an alternative that requires less in the way of behavioural and economic change becomes quite attractive – particularly to those with vested interested in the economic system remaining as it is including those with decision-making power in the areas of politics and policy making, business and innovation, academia and the press (Sikka 2013; Vidal 2012a, b; Panwar et al. 2011). Addressing its implications through a variety of lenses, while also pushing for sustainable growth models based on renewable energy, as such, has become of paramount importance.

Studies on and about geoengineering have been conducted with numerous reports produced by such respected scientific bodies as NASA, The National Research Council (NRC), and The Royal Society (The Royal Society 2009; NRC 2015; NASA 2016). In 2010, a joint report was issued for the U.S. House of Representatives Committee on Science and Technology and the United Kingdom House of Commons Science and Technology Committee after a number of hearings held by Congress and Parliament and extensive discussions by government agencies, scientists, academics, policy makers and other 'stakeholder groups.' Articles, both critical and optimistic of the process, have appeared in *The Guardian, The New York Times, Scientific American, Newsweek,* and *Bloomberg* (Fountain 2015a, b; Snyder-Beattie 2015; Venkataraman 2016; Rostom 2015).

Generally, the way in which geoengineering is treated is throughout the book is as an analytical category whose characteristics are imposed on a number of different technologies and processes with a shared end goal. It is these characteristics, assumptions and values that are philosophically, sociologically and politically meaningful. Discussing climate engineering in this way allows for engagement with the interests and background knowledge that inform these technologies and techniques without getting too mired in the scientific and technological minutiae – which is necessary for a project like this. These analytical characteristics are heuristically functional in the same way that Weberian ideal types are. For Weber, "An ideal type is formed by the one-sided accentuation of one or more points of view and by the synthesis of a great many diffuse, discrete, more or less present and occasionally absent concrete individual phenomena, which are arranged according to those one-sidedly emphasized viewpoints into a unified analytical construct" (Coser 1977, 223–224).

Although Weber's focus is on the ideal-typical characteristics of a modern society, it is possible to draw on this model in which geoengineering is seen as constituting a distinctive set of practices and techniques that are also ideal-typical. These characteristics reflect the technology's invasiveness, its large-scale application, generalized uncertainty, and globalizing reach. However, it is important not to impose static attributes onto geoengineering and to allow for new characteristics and meanings of these technologies, and the scientific practices that give rise to them, to emerge through processes of analysis and study. In order to analyse the portions of the book to be both consistent and manageable, I focus primarily on solar radiation management technologies (SRM) in the case study chapters using FCE with particular attention, paid to the use of stratospheric sulfate aerosols. Under this proposed geoengineering scheme, sulfate aerosols (either hydrogen sulfide or sulfur dioxide) are injected into the stratosphere in order to reflect harmful incoming radiation from the sun.

What makes geoengineering exceptional, as scientific and socio-technical artifacts, is that it is unites the natural sciences, physical sciences, computer modeling and prospective technological objects together. As such, a comprehensive gendered analysis requires that attention be paid to the ontology, background assumptions, values, and knowledge-gathering practices that comprise the mathematics, physics, chemical equations, eco-biological principles, representational assumptions, and technological choices that constitute climate engineering as a complex scientific and socio-technical phenomena. Feminist empiricism, with some support from standpoint theory and technofeminism, does an exceptional job in contending with the subject of values in climate science, the role of background assumptions, and the process of justification with respect to climate engineering.

Opposition to geoengineering has come from wide range of sources including scientists, academics, politicians, NGO's, and the media (Robock 2008a, b; Goes et al. 2011; Scientific American 2008; McLaren 2015). However, little work, within the domain of critique has been conducted from a feminist perspective. As such, I aim to pose considered objections to geoengineering on feminist grounds drawing primarily on a contextualist form of feminist empiricism which, as stated, is then supplemented with key insights from feminist standpoint theory and technofeminism.[1]

Feminism and Feminist Science

There are a variety of feminist approaches that can contribute to the study of geoengineering. These include more liberal feminist critiques that would highlight the representation, or severe lack there of, of women in both the upstream and downstream aspects of geoengineering, all the way to ecofeminist critiques of geoengineering as a technology and set of practices that are consistent with the hierarchal, exploitative, masculinist, and oppressive ethos of traditional Western science (Mies and Shiva 1993; Gaard 1993, 1997; Merchant 1980; Quinby 1990).

Nature, in this context, is seen

> ...as the excluded and devalued contrast of reason, includes the emotions, the body, the passions, animality, the primitive or uncivilised, the non-human world, matter, physicality and sense experience, as well as the sphere of faith, irrationality and madness" (Plumwood 1993, 19–20).

Ecofeminism offers a clear position on geoengineering in light of its central premise that the oppression of women and nature are interconnected and that "modern industrial societies...through their institutions, values, assumptions, beliefs, knowledge," operate in a way that "'explains, justifies, [rationalises] and maintains' relationships of domination and the subordination of women and nature" (Warren 1996: xii). Ecofeminism's primary adherents, including Vandana Shiva and Maria Mies (2014), have explicitly repudiated the use of geoengineering as beyond the pale arguing that,

> To now offer that same mindset as a solution is to not take seriously what Einstein said: that you can't solve the problems by using the same mindset that caused them. So, the idea of engineering is an idea of mastery (Shiva as cited by Anshelm and Hansson 2014a, b, 6).

Shiva and Mies analyze the commodification of nature and women's bodies along the lines of a patriarchy and exploitation as well as articulating a generalized critique of Western science as concealing "the impure relationship between knowledge and violence or force...by defining science as the sphere of a pure search for truth" along Enlightenment ideals (Mies and Shiva 1993, 46). In drawing a direct line between the devaluation of women, the exploitation of marginalized communities (along the lines of colonial injustice), and the damage inflicted on the environment by Enlightenment science, climate engineering would likely be seen by

---

[1]The reader might observe that I use climate engineering and geoengineering interchangeably throughout the text. This choice was made in line with how interchangeably these two terms are in the political, scientific, media and policy realms.

ecofeminists as consistent with ecologically destructive and masculinist history that has governed Western culture for centuries.

As such, while its role as a form of critique and activism against geoengineering is of significant importance, in the context of this book, it is limited in what it can add to feminist empiricism and its approach to climate engineering. Consequently, rather than focusing on liberal and ecofeminism, I have chosen to draw on frameworks that are rarely employed together in this context. I argue that Helen Longino's feminist contextual empiricism, when coupled with insights from technofeminism and feminist standpoint theory, provides a comprehensive and perceptive exploration of geoengineering as a gendered set of techno-scientific practices. This hybrid method requires a critical feminist analysis of geoengineering from upstream to downstream and include questions about its adoption as a plausible idea/solution to climate change and problematizes the science that supports and challenges it, the modeling that informs its possible outcomes, the way in which it is discursively justified and symbolically constructed, and the values and background assumptions that inform each of these layers of analysis. Overall, as stated, my overarching objective in this book is to provide a set of substantive arguments against geoengineering based on the feminist insights, principles, and values articulated primarily by FCE, but also FST and technofeminism.

Feminist Contextual Empiricism

Feminist Contextual Empiricism (or FCE) is the overarching framework used to pose a feminist critique of geoengineering in general and the deployment of stratospheric aerosols in particular. Its primary proponent is Helen Longino whose approach is taken as archetypical throughout this book. The choice of Helen Longino is based on her extensive work in the field of epistemology and her record of articulating a rigorous approach to science that incorporates contextual (social, political, cultural) values as well as constitutive values based on traditional socially accepted scientific practice (Longino 1990). Notably, Longino does this without lapsing into relativism by demonstrating that knowledge, while socially produced, is fundamentally distinct from personal opinion since objectivity must not only be rational (e.g. consistent with evidence), but also stand up to community assessment. Her approach is also pluralist in contending that a

> ...multiplicity of approaches that presently characterizes many areas of scientific investigation does not necessarily constitute a deficiency. As pluralists, we do not assume that the natural world cannot, in principle, be completely explained by a single tidy account; rather, we believe that whether it can be so explained is an open, empirical question (Longino et al. 2006, b).

Moreover, Longino's incorporation of gender and feminism into her framework lends itself to the this study by demonstrating how this mode of analysis can expose gender bias while articulating virtues that are grounded in a feminist practice and which are local, inclusive, contextual and social. In this way, Longino goes beyond what ecofeminism can do by avoiding the trappings of gender essentialism and focusing more on how "we in practice [can] do science as… feminist[s]" (Longino 2001a, b, 220).

On a general level, FCE takes as valid the basic assumptions of classical empiricism in which sensory experience (based on observation) forms the basis of knowledge, and an extant external world, which exists independent of us and of which we can gain knowledge, is believed to exist (Longino 1989, 1994a, b, 2001a, b; Anderson 1995a, b; Potter 2007). As such, FCE aims to deal directly with epistemological problems concerning "the nature of knowledge itself, epistemic agency, justification, objectivity and whether and how epistemology should be naturalized" (Potter and Alcoff 1993:1). The methods of obtaining knowledge as a practice under the FCE framework consists of inserting feminist insights into generalized empiricism and entails the following assumptions as articulated by Andrea Doucet and Natasha S. Mauthner: first, all scientific observations, conclusions and findings are value-laden and value judgements are made in every stage of investigation; second, feminist empiricism involves a theory of evidence based on sensorial observation; and third, knowers "are not individuals but are rather communities and more specifically science communities and epistemological communities" (Doucet and Mauthner 2006, 37).

FCE's conception of scientific practice is connected to this social model of knowledge formation and its view of science is rooted in theoretical virtues that are "characteristics of theories, models, or hypotheses" and "are taken as counting prima facie and ceteris paribus in favor of their acceptance." They are also consistent with what Longino maintains are virtues that feminist researchers would endorse including "empirical adequacy, novelty, ontological heterogeneity, complexity or mutuality of interaction of human needs, and decentralization of power or universal empowerment" (Longino and Lennon 1997a, b, c, 21).

The method through which I critique geoengineering using FCE is by subjecting the underlying science, modeling, assumptions, values and practices to each of these virtues. I demonstrate that, with a few exceptions, the socio-technical strategy that has come to be known as geoengineering does not meet the feminist criteria set out by FCE. The significance of this, and the reason why it matters, is that if the insights of feminist 'natural' science studies are to gain more traction, it is essential to use this approach to study science and innovation in practice. Moreover, the conclusions reached through this text will undoubtedly add a new dimension to the substantial literature that already exists which contends that climate engineering, as a solution to climate change, is a dangerous proposition.

Feminist Standpoint Theory

FST contributes much needed insights to FCE as it relates to the role(s) of situated knowledge, diverse epistemologies and power dynamics by refusing to "eliminate politics or interests from science," and, instead, focusing on,

...which interests advance knowledge and for whom, and which obstruct knowledge, again for whom. In other words, we always have to ask whose interests are served by the knowledge that mainstream science believes it is important to develop, and whose interests are served by the knowledge projects that are overlooked or ignored (Tuana 2013, 18).

Feminist standpoint theory is emphatic in its thesis that marginalized groups, including but not limited to women (and where women are not treated as monolithic), have significantly diverse perspectives and experiences that gives them a kind of epistemic privilege based on their familiarity with the social order wherein subordination and domination are constants (Harding 1991, 1993; Kourany 2009; Crasnow 2009). This positionality, according to Julia Seymour, results in a kind of strong objectivity in which the conclusions reached by women are "less distorted and self interested" and thus "will yield a more accurate picture of social reality" (Seymour 2013, 2).

Drawing on this feminist approach to study geoengineering requires an investigation of how existing research, models, approaches and conclusions would be different if gender were given a high priority and if women were given a voice. While both FST and FCE acknowledge the role history, culture, and political context play in knowledge formation, FST holds that any "approach to research and knowledge formation that does not acknowledge the role that power and social location [e.g. gender] play in the knowledge production process must be understood as offering only a weak form of objectivity" (Naples and Gurr 2013, 20). A more fulsome discussion of FST and what this mode of analysis contributes to a feminist critique of geoengineering that FCE does not is taken up in the conclusion.

Technofeminism

A particularly salient approach to gender, technology and science that provides a unique perspective on geoengineering is feminist technoscience. Feminist technoscientific theory, also known as technofeminism, combines feminism with science and technology studies in a manner that focuses on both the upstream and downstream aspects of technological development and use by positing that gender and technologies are co-constituted (Smelik and Lykke 2010; Wajcman 2004; Aschauer 1999). Judy Wacjman, who best articulates this approach, combines a generalized critique of technology with feminism to study ICTs in particular. Her thesis challenges the commonly held assumption that technological artifacts are simply the products of rational imperatives. Rather, Wajcman makes it clear that technologies must be seen as "sociotechnical product[s]" that combines "artefacts, people, organizations, cultural meanings and knowledge" (Wajcman, 7, 2009; Bijker et al. 1987; Law and Hassard 1999).

Although her focus is on media technologies, I argue that elements of Wajcman's work are applicable to geoengineering as well – particularly as it relates to how gender relations are implicated in technological development and use. Wacjman shows how unequal gender relations can influence innovation and design trajectories and maintains that,

> This points to the need to examine the ways in which the gendering of technology affects the entire life trajectory of an artefact. Integrating detailed studies of design, manufacture, purchase and consumption allows a range of social and technical factors, including gender to become apparent. For this reason, technofeminist approaches stress that gendering involves several dimensions, involving material, discursive and social elements (Wajcman 2007, 294)

Relevant issues with respect to geoengineering include how and in what ways this influence has been manifest. That is, how has the gendering of geoengineering been enacted on multiple levels including research, development, and use – which FCE does not account for. It also, as this quotation expresses, must include an examination of the discourse surrounding geoengineering as it has been deployed in media, scientific and policy circles. Technofeminism also deploys the unique contributions of domestication theory in which it is asked whether it is possible to draw on user agency in such a way as to transform technologies into a more prosocial, flexible, and democratic set of practices and techniques.

Breakdown

The way in which the book is organized is as follows: In Chap. 2, I provide an overview of geoengineering in general and the stratospheric sulphate SRM method in particular as one possible solution to the 'wicked problem' of climate change. I also continue the discussion of the relevance of FCE to these technologies. As stated, while climate engineering is treated as an analytic category or ideal type throughout much of this book, I focus specifically on the solar radiation management technique whose objective is to deploy radiation reflecting sulfur aerosols in the atmosphere.

In Chap. 3, I provide a detailed discussion of FCE and engage in a generalized critique of climate engineering using this method. Central to this chapter is a comprehensive investigation of the scientific assumptions and values that buttress geoengineering and the attendant norms, values and assumptions that underpin it with respect to computer modeling and simulation. The status of truth, scientific discovery, justification and pluralism is examined at length. Overall, FCE is used to critique, assess, inform and contest the values and norms that underlie geoengingeering as well as take on the role played by truth and justification.

In the following chapters, I undertake a thorough investigation of atmospheric sulphate geoengineering with respect to each of the virtues of feminist empiricism articulated by Longino. These include empirical adequacy, novelty, ontological heterogeneity, complexity or mutuality of interaction of human needs, and decentralization of power or universal empowerment. I argue that SRM sulphate geoengineering fails to meet the criteria of feminist science as expressed by these virtues and discuss how, why, and in what way this generates a compelling argument against the deployment of these technologies.

In the final Chapter I draw on FST and technofeminism to plug some of the gaps left by FCE in pursuit of a rounded feminist critique of climate engineering. In addition to discussing the role of situated and marginalized knowledge, FST is discussed as supplementing the feminist critique of geoengineering as it relates to the subject of power, which FCE does not engage in to any large extent. Technofeminism is also used to add to this analysis, specifically on the subjects of symbolic and discursive construction, as well as on whether, how, and in what way geoengineering can be understood as a gendered technology. Co- or mutual construction of gender and technology is examined as well as the role(s) of networks, political values, and context. The ability for technologies to develop in new ways is also considered. Technofeminism's central insight that "science and technology embody values, and have the potential to embody different values" (Wajcman 2004, 126), is of particular significance.

Methodologically, in each of these chapters I divide the feminist analysis of geoengineering into upstream and downstream elements. Upstream includes: the study of the process by which geoengineering has come to be constituted as a possible solution to climate change; the scientific assumptions, theories and principles that inform its socio-political construction; the role modeling and simulation plays in reaching conclusions about its feasibility, effects, and rates of success broadly defined; the role played by scientific pluralism; and the ways in which these techniques meet, or fail to meet, central feminist scientific virtues as articulated by Longino. Most of these layers of analysis at the upstream end are attended to sufficiently by FCE. However, I argue that FST adds indispensable upstream insights around representation, power, and multiple perspectives while technofeminism contributes significantly to the up or even mid-stream notion of mutual shaping (of technology and gender), as well as downstream aspects of symbolic and discursive construction and the interpretive flexibility of technology.

In formulating novel ways of interweaving gender into physical, environmental and computer science, it is important to be clear about why this matters and what it contributes. Generally, very little work has been done on how and in what ways the physical or natural sciences contain social values that are gender laden. Because gender is one of the fundamental categories that mark unequal power relations throughout society, it is important to fill this gap. As Sandra Harding notes, we must,

...understand knowledge-seeking as a fully social activity one that will inevitably reflect the conscious and unconscious social commitments of inquirers. From this perspective, it cannot be either merely accidental or irrelevant that most social studies of science, like their empiricist-guided ancestors, are loath to consider the effects on science of gendered identities and behaviors, institutionalized gender arrangements, and gender symbolism (Harding, 1987, 201).

It is my objective to provide an example of such a feminist analysis from the domain of climate change and geoengineering in this book.

References

Anderson, E. (1995a). Feminist epistemology: An interpretation and a defense. *Hypatia, 10*(3), 50–84.

Anderson, E. (1995b). Knowledge, human interests, and objectivity in feminist epistemology. *Philosophical Topics, 23*(2), 27–58.

Anshelm, J., & Hansson, A. (2014a). The last chance to save the planet? An analysis of the geoengineering advocacy discourse in the public sphere. *Environmental Humanities, 3*, 101–123.

Anshelm, J., & Hansson, A. (2014b). Battling Promethean dreams and Trojan horses: Revealing the critical discourses of geoengineering. *Energy Research & Social Science, 2*, 135–144.

Aschauer, A. B. (1999). Tinkering with technological skill: An examination of the gendered uses of technologies. *Computers and Composition, 16*, 7–23.

Bijker, W., Hughes, T., & Pint, T. (Eds.). (1987). *The social construction of technological systems.* Cambridge, MA: MIT Press.

Coser, L. (1977). *Masters of sociological thought: Ideas in historical and social context.* New York: Harcourt.

Crasnow, S. (2009). Is standpoint theory a resource for feminist epistemology? An introduction. *Hypatia, 24*(4), 189–192.

Doucet, A., & Mauthner, N. (2006). Feminist methodologies and epistemologies. In C. D. Bryant & D. L. Peck (Eds.), *Handbook of 21st century sociology* (pp. 36–42). Thousand Oaks: Sage.

Fountain, H. (2015a, February 10). Panel urges research on geoengineering as a tool against climate change. *The New York Times.* http://www.nytimes.com/2015/02/11/science/panel-urges-more-research-on-geoengineering-as-a-tool-against-climate-change.html. Accessed 15 Sept 2016.

Fountain, H. (2015b). Panel urges research on geoengineering as a tool against climate change. *New York Times*, 10 Feb 2015. https://www.nytimes.com/2015/02/11/science/panel-urges-more-research-on-geoengineering-as-a-tool-against-climate-change.html?_r=0. Accessed 24 Feb 2017.

Gaard, G. (1993). *Ecofeminism: Women, animals, nature.* Philadelphia: Temple University Press.

Gaard, G. (1997). Toward a Queer Ecofeminism. *Hypatia, 12*(1), 114–137.

Goes, M., Tuana, N., & Keller, K. (2011). The economics (or lack thereof) of aerosol geoengineering. *Climatic Change, 109*(3–4), 719–744.

Harding, S. (1987). *The science question in feminism.* Ithaca: Cornel University Press.

Harding, S. (1991). *Whose science? Whose knowledge? Thinking from women's lives.* Ithaca: Cornell University Press.

Harding, S. (Ed.). (1993). *The 'racial' economy of Science: Towards a democratic future.* Bloomington: Indiana University Press.

Kellert, S. H. (2006). Disciplinary pluralism for science studies. In S. H. Kellert, H. E. Longino, & C. K. Waters (Eds.), *Scientific pluralism* (pp. 215–230). Minneapolis: University of Minnesota Press.

Kourany, J. (2009). The place of standpoint theory in feminist science studies. *Hypatia, 24*(4), 209–218.

Law, J., & Hassard, J. (Eds.). (1999). *Actor-network theory and after.* Oxford: Blackwell.

Longino, H. E. (1990). *Science as social knowledge: Values and objectivity in scientific inquiry.* Princeton: Princeton University Press.

Longino, H. (1994a). In search of feminist epistemology. *Monist, 77*, 472–485.

Longino, H. (1994b). The fate of knowledge in social theories of science. In F. Schmitt (Ed.), *Socializing epistemology: The social dimensions of knowledge* (pp. 135–157). Lanham: Rowman and Littlefield Publishers Inc.

Longino, H. (2001a). *The fate of knowledge.* Princeton University Press: Princeton.

Longino, H. (2001b). Can there be a feminist science? In M. Wyer, M. Barbercheck, D. Giesman, H. Öztürk, & M. Wayne (Eds.), *Women, science, and technology* (pp. 216–222). New York: Routledge.

Longino, H. E., & Lennon, K. (1997a). Feminist epistemology as local epistemology. *Proceedings of the Aristotelian Society, Supplementary Volumes, 71*, 1–35.

Longino, H. E., & Lennon, K. (1997b). Feminist epistemology as a local epistemology. *Proceedings of the Aristotelian Society, Supplementary Volumes, 71*, 19–54.

Longino, H. E., & Lennon, K. (1997c). Feminist epistemology as local epistemology. *Proceedings of the Aristotelian Society, Supplementary Volumes, 71*, 19–35+37–54.

Longino, et al. (2006a). Introduction: The pluralist stance. In S. H. Kellert, H. E. Longino, & K. Waters (Eds.), *Scientific pluralism* (pp. vi–xxix). Minneapolis: Minnesota Press.

Longino, H., Kellert, S. H., & Waters, C. K. (2006b). *Scientific pluralism. Minnesota studies on the philosophy of science* (Vol. XIX). Minneapolis: University of Minnesota Press.

Luderer, G., et al. (2014). The role of renewable energy in climate stabilization: Results from the EMF27 scenarios. *Climatic change, 123*(3-4), 427–441.

McLaren, D. (2015, March 14). Where's the justice in geoengineering. *The Guardian.*https://www.theguardian.com/science/political-science/2015/mar/14/wheres-the-justice-in-geoengineering. Accessed 12 Sept 2016.

Merchant, C. (1980). *The death of nature - Women, ecology and the scientific revolution.* San Francisco: Harper Collins.

Mies, M., & Shiva, V. (1993). *Ecofeminism.* London: The Zed Press.

Naples, N., & Gurr, B. (2013). Feminist empiricism and standpoint theory: Approaches to understanding the social world. In S. Hesse-Biber (Ed.), *Feminist research practice: A primer* (pp. 14–41). Thousand Oaks: Sage.

NASA. (2016, July 14). Stratospheric aerosol and gas experimentation III-ISS (SAGE III-ISS). http://www.nasa.gov/mission_pages/station/research/experiments/1004.html. Accessed 22 Sept 2016.

NRC. (2015). *Climate intervention: Reflecting sunlight to cool the earth.* Washington, DC: The National Academies Press.

Panwar, N. L., et al. (2011). Role of renewable energy sources in environmental protection: A review. *Renewable and Sustainable Energy Reviews, 15*(3), 1513–1524.

Plumwood, V. (1993). *Feminism and the mastery over nature.* New York: Routledge.

Potter, E. (2007). Feminist epistemologies and women's Lives. In L. M. Alcoff & E. F. Kittay (Eds.), *The Blackwell guide to feminist philosophy* (pp. 235–253). Malden: Blackwell Publishing Ltd.

Potter, E., & Alcoff, L. (1993). *Feminist epistemologies.* New York: Routledge.

Quinby, L. (1990). Ecofeminism and the politics of resistance. In I. Diamond & G. Orenstein (Eds.), *Reweaving the world: The emergence of ecofeminism.* San Francisco: Berkeley Press.

Ren21. (2017). *Renewables: Global status report.* Renewable Energy Policy Network for the 21st Century. Available at: http://www.ren21.net/wp-content/uploads/2017/06/17-8399_GSR_2017_Full_Report_0621_Opt.pdf. Accessed 2 May 2018.

Robock, A. (2008a). 20 reasons why geoengineering may be a bad idea. *Bulletin of the Atomic Scientists, 64*(2), 14–18.

Robock, A. (2008b). Whither geoengineering? *SCIENCE-NEW YORK THEN WASHINGTON, 320*(5880), 1166.

Rostom, E. (2015, February 10). *Geoengineering: The bad idea we need to stop climate change.* Bloomberg. http://www.bloomberg.com/news/articles/2015-02-10/geoengineering-the-bad-idea-we-need-to-stop-climate-change. Accessed 17 Sept 2016.

Scientific American. (2008). The hidden dangers of geoengineering. *Scientific American.* Available at: https://www.scientificamerican.com/article/the-hidden-dangers-of-geoengineering/. Accessed 2 Jan 2017.

Seymour, J. (2013). Feminist standpoint epistemology – The role of women in climate change policy-making: Are some people's experiences more valuable than others' as a foundation for knowledge and generating social change? *Pragmatism Tomorrow, 1*(3), 1–10.

Shiva, V., & Mies, M. (2014). *Ecofeminism.* London: Zed Books Ltd..

Sikka, T. (2013). An analysis of the connection between climate change, technological solutions and potential disaster management: The contribution of geoengineering research. In W. Filho (Ed.), *Climate change and disaster risk management* (pp. 535–551). Berlin: Springer.

Smelik, A. M., & Lykke, N. (Eds.). (2010). *Bits of life: Feminism at the intersections of media, bioscience, and technology*. Seattle: University of Washington Press.

Snyder-Beattie, A. (2015, May 15). Geoengineering is fast and cheap, but not the key to stopping climate change. *The Guardian*.https://www.theguardian.com/sustainable-business/2015/may/15/geoengineering-climate-change-greenhouse-gases. Accessed 17 Sept 2016.

The Royal Society. (2009, September). *Geoengineering the climate: Science, governance and uncertainty*. London: Royal Society of London. https://royalsociety.org/topics-policy/publications/2009/geoengineering-climate/. Accessed 17 Sept 2016.

Tuana, N. (2013). Gendering climate knowledge for justice: Catalyzing a new research agenda. In M. Alston & K. Whittenbury (Eds.), *Research, action and policy: Addressing the gendered impacts of climate change* (pp. 17–31). Dordrecht: Springer.

Twidell, J., & Weir, T. (2015). *Renewable energy resources*. New York: Routledge.

Venkataraman, B. (2016, January 14). We have no way to predict the unintended consequences of geoengineering. *Newsweek*.http://www.newsweek.com/geoengineering-unintended-consequences-blocking-sun-415449. Accessed 15 Sept 2016.

Vidal, J. (2012a). Bill Gates backs climate scientists lobbying for large-scale geoengineering. *The Guardian*, 6 February 2012. https://www.theguardian.com/environment/2012/feb/06/bill-gates-climate-scientists-geoengineering. Accessed 13 Feb 2017.

Vidal, J. (2012b, February 6) Bill Gates backs climate scientists lobbying for large-scale geoengineering. *The Guardian*.https://www.theguardian.com/environment/2012/feb/06/bill-gates-climate-scientists-geoengineering. Accessed 4 Oct 2012.

Wajcman, J. (2004). *Technofeminism*. Cambridge/Maiden: Polity Press.

Wajcman, J. (2007). From women and technology to gendered technoscience. *Information, Community and Society, 10*(3), 287–298.

Wajcman, J. (2009). Feminist theories of technology. *Cambridge Journal of Economics*, 1–10.

Warren, K. J. (1996). *Ecological feminist philosophies: An overview of the issues*. Bloomington: Indiana University Press.

World Energy Council. (2016). *World Energy Resources*. World Energy Council. https://www.worldenergy.org/wp-content/uploads/2016/10/World-Energy-Resources-Full-report-2016.10.03.pdf. Accessed 1 May 2018.

Chapter 2
Geoengineering

Abstract This Chapter provides an overview of geoengineering research, including the status of current research and testing, the significance of modeling and simulation, the role of public participation, and the subject of governance. A discussion of geoengineering's basic epistemology, values and background assumptions is also included with specific attention paid to the solar radiation management technique of atmospheric sulfate geoengineering.

Keywords Atmospheric sulphate geoengineering, · Feminist empiricism · Helen Longino · Solar radiation management · Pluralism · Boundary object · AOGCM

This Chapter provides an in-depth overview of geoengineering research as an ideal-typical technoscientific set of practices. It's purpose is to provide the reader with a basic understanding of what geoengineering entails, the status of current research and testing (including a mapping of the central actors), the significance of modeling and simulation, the role of public participation, and the contentious subject of governance. A discussion of its epistemology, values and background assumptions are also included – which are particularly significant as it relates to subsequent Chapters on feminism and feminist science. Engaging in a feminist critique of geoengineering requires substantive knowledge of all of these factors to properly investigate the scientific practices and assumptions that constitute geoengineering, the place and role of marginalized perspectives, and the gendered nature of the technology itself.

This Chapter also provides an introduction into the SRM technique of atmospheric sulfate seeding. It is important to explain the rationale, scientific complexities, current research status, and general standing this SRM technique in preparation for its use as a case study further on. Specifically, these sections will set the stage for the complex assessment of this particular geoengineering technique against Helen Longino's feminist scientific virtues. This analysis aims to introduce a novel and urgent gendered dimension to the critique of geoengineering and to arguments against its future use.

Climate Change: Setting the Stage

The effects of global warming on the world's climate have been established with robust evidence and vigorous scientific consensus (Anderegg 2010; Doran and Zimmeman 2009; Cook et al. 2016). Yet actions to address its impacts, which range from rising sea levels and uncertain precipitation patterns to more intense heat waves and droughts, that have been entirely insufficient. A further warming of 1.5–2 °C is seen as a boundary that, if crossed, will lead to serious planetary consequences. Proposals from greater energy efficiency and energy alternatives (solar, wind, geothermal) to limiting carbon emissions and/or monetizing pollution via carbon markets, trading and taxes constitute the more widely cited solutions to climate change. Yet, what has become patently clear is that, as it stands, we are likely to exceed these emissions limits since, as Bruns and Strauss argue, "on balance the political will to make the necessary effort to reduce carbon emissions does not exist. What is more, it does not seem likely to come about within the time frame necessary to stave off very serious conclusions" (Bruns and Strauss 2013, 2).

Against this backdrop, it is useful to draw on the theory of 'wicked problems' to understand the unique set of issues climate change poses and how geoengineering has come to be seen as a possible solution. The term wicked problem gained popularity in the 1970s with respect to public planning. Wicked problems pose a unique sets of challenges for policy makers, political actors and the general public as a result of their entanglement with other complex problems and lack of clarifying traits including: a definitive formulation, a clear endpoint, an ability to learn by trial and error, and a solution that can be judged as true or false (Rittel 1973). As such, their inherent complexity makes it difficult to pose dispositive solutions and garner agreement by what are often self-interested actors.

Environmental wicked problem in particular pose intractable barriers to clear resolutions. They are characterized by a sense of urgency (e,g. 'a small window'), scientific 'messiness', conflicting risks, dynamic social, economic, knowledge and technological systems, and the lack of a central authority with the power to enforce a decision (Balint et al. 2011, 14). On the subject of climate change, the tension between urgency, risk, and complex social systems is particularly profound. The science is also, while consensus based, subject to the 'messiness' of multiple variables, political uncertainty, model uncertainty and system variability (Balint et al. 2011, 18019). Moreover, while the establishment of bodies like the Intergovernmental Panel on Climate Change (IPCC) and the Paris Treaty are laudable steps towards some sort of central authority, they lack the power to enforce cuts in emissions.

Because it is a quintessentially wicked environmental problem, and geoengineering is increasingly posed as a possible solution to climate change. Specifically, it is presented as a kind of plan B if mitigation, enacted through cuts in GHG emissions, the adoption of cleaner renewable energy, generalized energy efficiency and decarbonization, and adaptation (which integrates "improved early warning systems, hardening of infrastructure, particularly in coast areas…improved water management…development of drought- and salt-tolerant agricultural grains, and procedures for dealing with relocation of large human population" (Incorpera 2016, 250–251)), fails to be sufficient. This potential failure, coupled with the sense that a catastrophe is imminent and an overall desire not to have to change our socio-economic and consumption based order in any substantive way, has made climate engineering a more palatable and even desirable option.

Several researchers have begun to call geoengineering a wicked problem in and of itself – particularly in light of its distinctive scientific, economic, policy-driven, and ethical complexities. Zhang and Posch articulate six major arguments in support of the proposition that geoengineering poses a 'wicked problem,' including "complex cross-boundary feedbacks, economic affordability, decision-making criteria, conflicting interests and values, lack of central governance, and tuxedo fallacy of decision making" (Zhang and Posch 2014, 393). The first five of these questions are taken in detail up throughout this Chapter, while the last question, the tuxedo fallacy, simply refers to the tendency to treat decision making as if one had reliable estimates of the probability of all possible results (Hansson 2008; Zhang and Posch 2014). With respect to geoengineering, this paradox reveals that, it is not possible to anticipate probable outcomes without actually executing geoengineering itself.

Geoengineering

As stated, geoengineering is treated in this book as an ideal-typical or analytic category that is constituted by several shared characteristics including: its objective or intent to counter and/or mitigate climate change technologically; its scale, which is global and transformative; and the degree to which the plan or scheme is seen as a countervailing measure (Keith 2000, 247). As a countervailing measure, climate engineering techniques are often perceived as potential magic bullets with the capacity to solve the climate crisis.

Most variants of climate these technologies are also colloquially framed as apolitical. This is consistent with a Lockean and modernist understanding of science in which it, and its technological applications, are seen as neutral, objective, unbiased, and qualitatively "different from all other kinds of human knowledge since it can be verified and therefore can be said to be universally true" (Delanty 1997, 12). This is despite of the fact that this understanding of science has and continues to be challenged by scholars, including feminists (Feenberg 1999; Winner 1980; Harding 1991). As such, political solutions involving treaties, emission cuts, changes in consumption habits, economic restructuring, and new ways of approaching our relationship with the natural world are seen as regressive, unreliable, and value-laden. I return to this theme in subsequent Chapters.

Brief History

Geoengineering has a significant and informative history worth noting in brief. It's roots can be traced to the 1960s and 1970s and, in particular, the US military's interest in weather modification as a way to gain military advantage. Schemes that included seeding clouds with chemicals to encourage rain were of interest to farmers as well as the military. In the 1960s, a particular military project, Project Stormfury,which was a joint venture between the US Navy and the US Weather Bureau, aimed to seed supercooled water in the eye of an impending hurricane in order to mitigate its impact (Willoughby et al. 1985). Globally, Russia has been known to have weather modification programs aimed at reducing drought and tornados while China has weather modification programs in use today (Lou et al. 2012; Taubenfeld and Taubenfeld 2009). With respect to war, during the Vietnam War sorties were flown by the US military to seed clouds in order to extend the monsoon season for military advantage (Harper 2008). Reagan's so called 'Star Wars' or Strategic Defense Initiative (SDI) project of space-based missile defense also contained a project aimed at disrupting weather patterns (Fleming 2007).

Prior to being formally brought under the auspices of the National Science Foundation, projects like these were often termed weather modification and only began to resemble what we recognize as geoengineering in the 1990s. Their early incarnations were banned in 1977 when the Convention on the Prohibition of Military or Any Other Hostile use of Environmental Modification Techniques was signed (Fleming 2006). It was not until the 2006 article in *Climate Change* by Nobel Prize winning scientist Paul Crutzen that geoengineering began to be taken seriously. In it, Crutzen argued that further research into climate engineering is needed as well as arguing specifically for the SRM reflective aerosol approach (Crutzen 2006).

Since then, there has been a surge in geoengineering research in the realms of science, modeling, and policy (Bellamy et al. 2012; Blackstock 2012; Cairns 2014; Blain et al. 2008; Stolaroff et al. 2008; Ammann et al. 2010). With respect to modeling, which is taken up further on (in addition to science and policy), significant movements have been made on dynamic features (ocean and atmosphere), and basic physics. These newer "Earth System" models incorporate "previously neglected processes such as atmospheric chemistry, aerosols, the terrestrial biosphere change and ocean biogeochemistry" (Lunt 2013, 18). The introduction of multiple models and comparative projects like the Geoengineering Modeling Intercomparison Project are examined extensively in the section on modeling (Kravitz et al. 2011). My use of the Kravitz et al. text as emblematic or representative of similar trends and results in other attempts at modeling geoengineering outcomes is a result of its incorporation of multiple experiments, being an intercomparison project, which, as Kravitz et al. asserts, provides a way to "compare results and determine the robustness of their responses" (Kravitz et al. 2011, 162), in a manner not present in other attempts at modeling (Kalidindi et al. 2015; Niemeier et al. 2013).

Climate engineering's acceptance as a viable area of study by mainstream scientific bodies and universities has risen significantly. The Royal Society, for example, supports further research (2009), while The American Meteorological Society's stated position is that "Geoengineering could conceivably offer targeted and fast-acting options to reduce acute climate impacts and provide strategies of last resort if abrupt, catastrophic, or otherwise unacceptable climate change impacts become unavoidable by other means" (American Meteorological Society 2009). The Max Planck Institute, UK Met Office and Carnegie Institution have all produced a publications on this subject (Jones et al. 2009; Matthews and Caldeira 2007; Oldham et al. 2014).

Definitions

The most straightforward and widely cited set of definitions of geoengineering I have come across is from the Oxford Geoengineering Programme, which has been housed in the Oxford Martin School at the University of Oxford since 2010. Their definitions are based on those articulated by The Royal Society in 2009. In its most general sense, geo- or climate engineering can be understood as set of techniques, technologies, and scientific programs that engage in "the deliberate large-scale intervention in the Earth's natural systems to counteract climate change" (Oxford Geoengincering Programme 2016). Geoengineering is itself constituted by two broad categories. The first is Solar Radiation Management (SRM) whose objective is to reflect sunlight away from the earth, while the second, CDR or carbon dioxide removal, aims to directly remove CO_2 emissions from the atmosphere and sequester it.

SRM consists of such methods as: albedo enhancement, which aims to increase the reflectivity of clouds by, for example, increasing the number of water droplets inside clouds to make them brighter and more reflective; the use of stratospheric aerosols to reflect sunlight in the upper atmosphere in a manner that mimics the effects of volcanic eruptions (see Fig. 2.1); and the use of reflective surfaces to block the sun's rays (such as painting rooftops white or positioning reflectors in space). David Keith argues that the central idea behind SRM is to "make the planet a little more reflective" which, in turn, will "partially and imperfectly compensate for the buildup of greenhouse gases like carbon dioxide, which are tending to trap heat and make the earth warmer" (Keith 2015).

SRM approaches are often seen as having immediate effect (see Matthews and Caldeira 2007) and are thought to be relatively inexpensive (McClellan et al. 2010). Yet due to criticisms that it deals with the symptoms rather than the root causes of climate change – coupled with uncertainty around potential side effects that include reduced precipitation and monsoon disruption (Ricke et al. 2010), ambiguity around termination effects, whitening of the sky, and ozone depletion (Robock et al. 2008; Solomon and Oceanic 1999) – taking steps to deploy such technologies is believed to be dangerous (Kintisch 2012).

The second category, Carbon Dioxide Removal (CDR), seeks to attack the underlying causes of climate change by removing carbon dioxide from the atmosphere directly. It includes a variety of different techniques such as enhanced weathering, which exposes "large quantities of minerals that will react with carbon dioxide in

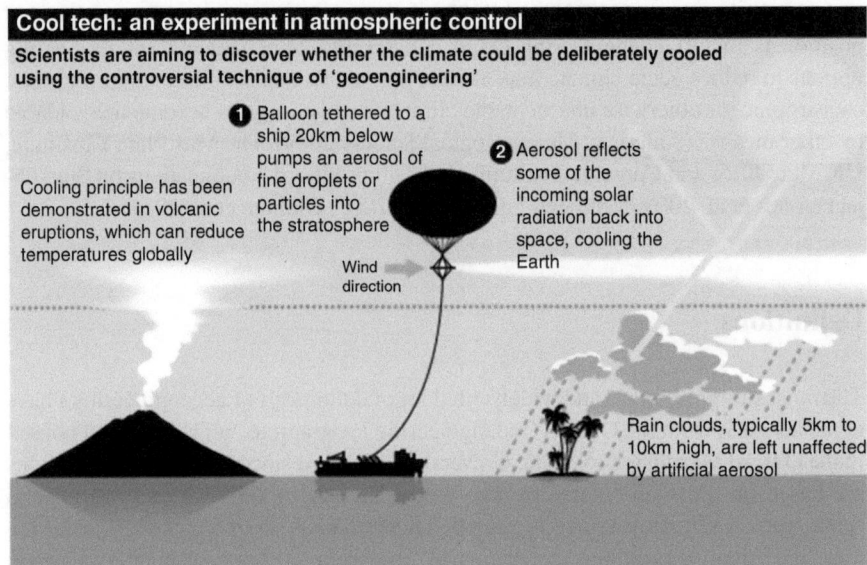

Fig. 2.1 Various forms of geoengineering (National University of Singapore Blog 2015)

the atmosphere and stor[e] the resulting compound in the ocean or soil" (Oxford Geoengineering Programme 2016); biochar, which burns biomass through pyrolysis and sequesters the carbon; and ocean fertilization which is another increasingly popular CDR geoengineering approach that entails intentionally "adding nutrients [like iron] to the ocean in selected locations to increase primary production [of algae] which draws down carbon dioxide from the atmosphere" (Oxford Geoengineering Programme 2016) (see Fig. 2.2).

CDR techniques are less popular because of their tendency to be rather slow in affecting substantive change to the climate and their cost – which is estimated to be more expensive than switching to non-carbon energy alternatives. Others contend that the overall complexity of the earth's systems, uncertainty involved in sequestration safety (Gardiner 2010), oceanic flux, and attribution problems constitute compelling reasons why CDR is a less than desirable option. For instance, with respect to attribution, in the case of iron fertilization, the traditional use of satellites to monitor changes to chlorophyll (which indicate growing algae blooms) have been found to be limited in its ability to identify the type of plankton in each bloom or the amount of carbon re-released (Abate 2013, 230; Dean 2009, 328) Yet CDR is also thought to be more effective with fewer potential side effects overall when compared to SRM (Kintisch 2012).

There is also a small subset of climate engineering termed 'natural geoengineering,' which aims at expanding our ecosystem's own capacity to store carbon and is often posed in opposition to the more aggressive geoengineering stratagems noted above. Reforestation or afforestation and biochar are two such examples. Afforestation or reforestation proposes to plant more carbon-removing trees in

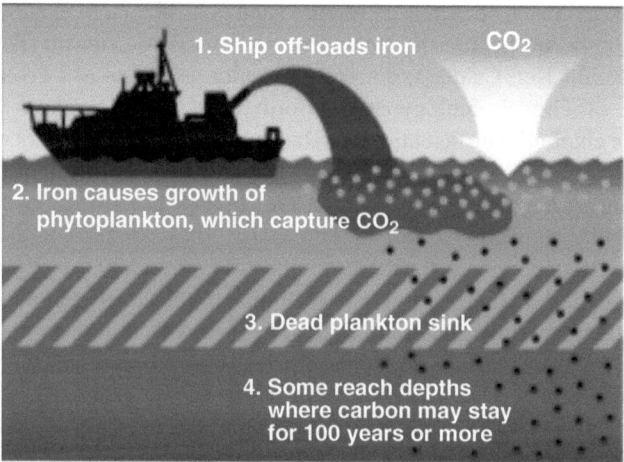

Fig. 2.2 Iron fertilization geoengineering (Geoengineering 2012)

previously cleared areas. While its results are not immediate, this mode of climate change mitigation has wide support. Biochar, as noted, applies a process of burning biomass (e.g. agricultural and municipal waste) and combining the resultant nutrient and carbon rich charcoal with soil. Another quite novel proposal that blurs the line between geoengineering and mitigation is the use of natural predators to limit the overpopulation of species that tend to overgraze on carbon absorbing grass and, in turn, reduce the production of methane and ammonia rich animal dung. This has worked well in the Serengeti which, through this approach, has "reverted to a carbon sink so large that it is estimated to offset all of East Africa's current annual fossil fuel carbon emissions" (Schmitz 2016a, b). I argue, however, that these natural approaches do not meet the levels of invasiveness and scale of more mainstream goengineering and are thus not examined further. '

Actors and Status of Research

It is enormously difficult to try and map out parties currently involved in geoengineering science, research, policy, advocacy and activism because of the sheer depth of the field. Most work in in the realm of geoengineering science and technology, with a few exceptions, are fixed in a research and modeling phase with very little in the way of field tests. Adding to this difficulty, both supporters and detractors of climate engineering differ in their levels of support or opposition, backing of particular techniques over others (e.g. SRM versus CDR), and general levels of influence. Moreover, they encompass a range of individual actors in the scientific community as well as governments, private citizens, NGOs, think tanks, and independent advocacy groups.

Scientific advocates of increased research and possible deployment in the future include atmospheric chemist and Nobel Prize winner Paul Crutzen; Harvard physicist David Keith; Ken Caldeira, currently a senior scientists and Professor in the Department of Global Ecology at Stanford; the Pacific Northwest's National Laboratory's chief climate scientist Phil Rasch; Marcia McNutt, an oceanographer who chaired the NAS study of geoengineering and is the editor-in-chief of *Science* magazine; and Granger Morgan, currently the Head of Department of Engineering and Public Policy at Carnegie Mellon University. They comprise a just handful of many scientific supporters of further geoengineering research.

It is important, however, to point out that most of the advocates of climate engineering in the scientific/academic community are what I like to call 'reluctant defenders' who clearly acknowledge the risks associated with adopting these technologies and practices. David Keith and Ken Caldeira, for example, argue convincingly that efforts to mitigate climate change should focus on so-called traditional strategies that includes internationally coordinated cuts in emissions and investment in, and transitional towards, sustainable energy technologies like wind, solar and geothermal. On David Keith's, Keith Group website, which is a collaborative research institute at Harvard University that examines the "intersection of climate science and technology with a focus on the science and public policy of solar geoengineering," it is made clear that these technologies could and should "not be a replacement for reducing emissions (mitigation) or coping with a changing climate (adaptation)" but rather, contends that "it could supplement these efforts" (The Keith Group 2016).

With respect to scientific bodies, the Royal Society and NAS (the latter of whose findings were supported by the CIA, NOAA and NASA) have conducted studies and given support for further research. The IPCC, after a thorough review of geoengineering in their 2013 AR5 Summary Report (IPCC 2013), included a statement on climate engineering's possibilities and drawbacks while the WMO, whose focus is on weather modification, officially stated that geoengineering should be studied for potential future use with clear parameters around governance and a comprehensive understanding of moral, natural, and political consequences (WMO 2013). Additionally, as noted, both the UK and US governments, through their Congressional and Parliamentary scientific committees, have published reports on geoengineering. The American Government Accountability Office (GAO) also issued a study into SRM and CDR in which it concluded that, these technologies are nowhere near ready for deployment, calling most measures "immature, many with potentially negative consequences but further research should be conducted" (GAO 2010).

Also worth mentioning is the 2010 Asilomar Conference On Climate Intervention Technologies in California which gathered 175 experts to come up with guidelines for conducting further research. Their basic conclusions are in line with the Oxford Principles on governance, but also discussed were issues around inequality, risk, the status of field experiments, and the pursuit of profit. The final report also reaffirmed the need to continue efforts at mitigation, but underlined that geoengineering may be necessary if this proves insufficient (Asilomar 2010).

Additionally, rich billionaire philanthropists have also gotten into this area primarily via financial support and publicity. Bill Gates, for example, has financed geoengineering research. In 2012, he gave millions to a fund managed by David Keith to research, and potentially test, SRM sulphate climate engineering. However, Gates has also continues to fund and invest in renewable energy R&D. Sir Richard Branson and Skype co-founder Niklas Zennstrom are also actively funding their own geoengineering projects and research (Marshall 2014). Many of the top ten finalists of Branson's Virgin Earth Challenge contest were geoengineering-related including projects for carbon capture and enhanced weathering (Branson 2016).

On the corporate end, we have companies like Climos – an iron seeding company involved in for-profit iron fertilization. Its head is tech tycoon, Dan Whaley, has gained both funding and support from entrepreneur Elon Musk. Global Thermostat is another company working on the future use of CDR techniques and is currently developing a machine that would extract CO_2 from the atmosphere – for a cost. A final actor of note is Russ George, an American businessman and active supporter of CDR geoengineering who, in 2012, dumped 100 tons of iron into the ocean off of Haida Gwaii in British Columbia in violation of national and international anti-dumping laws. The project was conducted in conjunction with the local Aboriginal community in the hope that it would boost salmon populations. George's vocal support of climate engineering, and the mandate of his now defunct company, Planktos, suggests that the test was much more about climate engineering than salmon restoration (Specter 2012a, b; Lucaks 2012).

Conservative think tanks like The Cato Institute, The Hudson Institute, and The American Enterprise Institute (AEI) are also prominent actors in geoengineering research. AEI vigorously supports further investigation into climate engineering – which is curious since their public stance on human induced climate change is that it is not scientifically proven (Dunlap and Jacques 2013). Yet, on geoengineering their stance is clear. AEI calls it a perfect,

> ...marriage of capitalism and climate remediation...What if corporations shoulder more costs and lead the technological charge, all for a huge potential payoff?...Let's hope we are unleashing enlightened capitalist forces that just might drive the kind of technological innovation necessary to genuinely tackle climate change (American Enterprise Institute as cited by Smoker 2013).

AEI's 'Geoengineering Project' is an interesting mixture of this kind of climate denialism coupled with optimism about geoengineering and presented in such a way as to accept the need for small reductions in carbon use and 'address' the problem of climate change via profit-driven technocapitalism. It is also noteworthy that corporations like Exxon and BP provide a significant amount of funding to these think tanks with which they are ideologically aligned (Hamilton 2013).

Opponents

Notable critics of climate engineering include Jane Long of the Lawrence Livermore National Laboratory in the US, Clive Hamilton, Professor of Public Ethics at the Australian National University, Alan Robock, Professor of Environmental Sciences at Rutgers University, and Raymond T. Pierrehumbert, who is the Halley Professor of Physics at the University of Oxford. Alan Robock is particularly active in this area. His article, "20 reasons why geoengineering may be a bad idea," articulates a persuasive set of arguments against geoengineering from an ecological, ethical and practical perspective. He asserts that possible risks emanating from its deployment include the possibility of ozone depletion, the likely disruption of regional climate, human error, high cost, an inability to 'go back,' the question of who decides, the undermining of mitigation, and possible militarization. Robock argues that, "If global warming is a political problem more than it is a technical problem, it follows that we don't need geoengineering to solve it" (Robock 2008a, b, 18).

The issues of moral hazard and path dependency are also cited as posing significant problems by opponents if governments and publics reduce their commitment to cut fossil fuel consumption in the belief that geoengineering technologies will 'fix' climate change. In their critiques of geoengineering, Stephen Gardiner and Clive Hamilton explain the moral hazard argument which contends that engaging in actionable climate engineering research represents a form of moral corruption in which Plan B (geoengineering) is chosen even through it represents a riskier and potentially less effective option than Plan A (mitigation through cuts to carbon emissions and the use of alternative energy). Hamilton argued that by choosing B instead of A out of self-interest, "we succumb to moral failure" (Hamilton 2011, 4; Gardiner 2010). With respect to path dependency and lock-in, several scholars have warned about the very real possibility that seemingly benign choices that permit further geoengineering research can constrain future choices and increase the chances that it will be used (Rayner et al. 2013; Shepherd et al. 2009; Cairns 2014). This can occur both cognitively, by helping to set the terms of early debate, and technologically wherein sociotechnical relations congeal behind a particular path forward (Healey 2014, 11; Bowker 1992).

In addition to individual academics, there are also several activist groups – many of whom have a strong online presence –who oppose geoengineering. They include the respected Canadian based ETC Group which is an activist organization explicitly opposed to open air testing of geoengneering. ETC runs a website titled "Geoengineering Our Climate" dedicated to challenging "geoengineering and other false solutions to climate change (e.g., proprietary, genetically-engineered 'climate-ready' crops)" and supporting "peasant-led agroecological responses to the climate crisis" (ETC Group 2015). Greenpeace, BiofuelWatch, Third World Network, The African Center for Biosafety and the German NGO Forum have also voiced opposition to the deployment of these technologies. Greenpeace scientist Doug Parr, for example, asserts that, "tinkering with our entire planetary system is not a dynamic new technological and scientific frontier, but an expression of political despair"

(Parr as cited by Specter 2012a, b). Parr is especially opposed to the code of conduct governance option discussed below. He asserted that, "scientists are not the best people to deal with the social, ethical or political issues that geoengineering raises… The idea that a self-selected group should have so much influence is bizarre" (Parr as cited by Vida 2012a, b).

In a happy but all too infrequent turn of events, activist organizing in the Global South have also carved out a presence on this subject. For instance, in 2010, the World's Peoples' Conference on Climate Change and the Rights of Mother Earth held in Cochabamba, Bolivia, launched a campaign called HOME (Hands Off Mother Earth) to oppose all geoengineering experimentation (World's Peoples' Conference on Climate Change and the Rights of Mother Earth 2011). The ETC Group is a founding member of HOME as they are Campesino groups and Indigenous organizations. Noted Indian physicist, environmentalist and ecofeminist scholar Vandana Shiva has also been vocal in her opposition to geoengineering calling it "the ultimate hubris without democratic control" (Heibel 2013). Finally, The African Biodiversity Network (ABN) has done grassroots work on geoengineering – most recently on land rights and biochar with respect to the threat it poses to biodiversity (ABN et al. 2009).

While the theme of representation is taken up in more detail in the Chapters on FST, it is worth noting that, there is a striking lack of women, represented in each of these categories. There is also insufficient representation of people of color or other minority groups at all levels. Moreover, the majority of the scientific research projects and attendant conversations are conducted in Western countries (North America, Western Europe etc.). There are, however, several initiatives underway to address this lacunae including the 2013–2014 Solar Radiation Management Governance Initiative (SRMGI) series of conferences and outreach meetings held in selected developing countries including India, Senegal, South Africa, Pakistan and China. However, it is important to note that the subject of these meetings was on governance, not upstream research. This indicates a troubling imbalance with respect to resources, power, authority and voice. There are, however, a handful of exceptions. In the section on SRM geoengineering, for example, I outline current work being done in China, which has undertaken significant scientific research on geoengineering. Russia is another country of note that has pushed, and conducted, extensive basic geoengineering research and who were vocal advocates of incorporating geoengineering into the IPCC AR5 report. Nevertheless, overall, with respect to the countries that make up the Global South, their voices are often absent from these conversations.

Public Participation

The role of public participation and deliberation is of tremendous importance not only to responsible science, but also to developing a responsive feminist understanding, analysis, and critique of geoengineering. Public participation in scientific research has become an important part of scientific practice as 'citizen science' (Bonney 1996) and collaborative projects are seen as both desirable and necessary. This is particularly the case in light of scientific and technological endeavors like genetic engineering, nanotechnologies and geoengineering which have tangible effects on the public at large. Science that falls into these categories can be categorized as examples of 'Post-Normal scientific practice' wherein the traditional separation between science, ethics, politics, and society are no longer relevant (Gibbons 1999; Sarewitz 2010).

Practicably, it is difficult to imagine a situation in which substantive participation on a global level in decision-making on geoengineering is logistically possible. Yet, as stated, it is widely accepted that some level of public involvement is necessary with respect to both a "recognition of basic human rights regarding democracy and procedural justice…" as well as "from a practical recognition that implementing unpopular policies may result in widespread protest and reduced trust in governing bodies" (Rower and Frewer 2000, 5). Modes of participation include an array of options such as collaborative forums which involve informing and briefing participates on the science, facilitated discussions. Q and A meetings with experts, and documented recommendations (Connick and Innes 2003; Seidenfeld 2000),

Burns and Flegel (2015) outline other modes of participation including consensus conferences, whose objective is to attain consensus on relevant questions and recommendations; a more global approach wherein participants at multiple sites engage in deliberation on particular themes and whose conclusions and questions are then put online; and deliberative mapping, in which participants, having been interviewed using software, are given options they must assign performance scores to. Participants then rank these options in order to foster "reflexive deliberation by not imposing pre-ordained definitions or weightings on criteria" (Burns and Flegel 2015, 277).

Burns and Flegel also summarize non-deliberative mechanisms including online surveys, polls and focus groups. In all of these options, issues of power, representation (with respect to gender, race, and class, individual versus group autonomy, and special interests) figure prominently. Feminist approaches to science and technology also tend to support public participation since it emphasizes research that is collaborative ,

> …not exploitative… It is, of necessity, knowingly embedded in a social and political context, which is taken into explicit account in the research process and reflective practice. Interactions with research collaborators lead participatory researchers to assess their categories and assumptions critically and mutually. [italics in source] (Fortmann et al. 2008, 82)

As a result, there have been a variety of proposals and studies aimed at engaging active publics and suggesting ways in which civil society perspectives can be incorporated into both upstream and downstream climate engineering research.

A particularly salient example of the institutional acknowledgment of the importance of public participation appears in an NAS study of climate engineering. In February 2015, the NAS released two reports on geoengineering (2015a, b), one on CDR and the other on SRM respectively, which take up the subject of governance in general and public participation in particular. Overall, the NAS is more favorably inclined to the use CDR techniques – which they consider to be precise and controllable. Yet the report concludes that because of the seriousness of potential risks,

> If a new governance structure is determined to needed...the governance structure should consider the importance of being transparent and having input from a broad set of stakeholders and appropriate consideration of all dimensions (NAS, 156, 2015a; NAS, 10, 2015b).

However, it important to note that it is often the case that the civil society groups most often engaged with by scientists and governmental groups in exercises of this sort tend to be large and institutionalized NGOs with little direct participation from the general public.

Nevertheless, significant, albeit selective, acts of engagement aimed at ascertaining public perceptions of climate engineering have been conducted (Parkhill and Pidgeon 2011; Ipsos-MORI 2010; Mercer et al. 2011). In a 2012 study, authors Nick Pidgeon et al. conducted a series of interviews and surveys to gage British public opinion on geoengineering. The study found very low levels of awareness, with 75% unaware of geoengineering, overall support for these technologies, a strong correlation between formal education and knowledge of geoengineering, and a positive relationship between those who believed climate change is caused by humans and support for SRM and CDR (with a preference for the latter) (Pidgeon et al. 2012). In a German study of 'layperson's' opinions on climate engineering, the authors found that most participants support geoengineering as a backup strategy with a preference for lower risk strategies such as the cloud whitening technique (Amelung and Funke 2014).

Also important, with regard to public participation, is the state of public discourse on geoengineering as expressed by the media. In an examination of articles from *The Guardian* and *The New York Times*, Luokkanen et al. (2013) found that arguments for and against geoengineering shared common frames. Those that support its use tend to draw on insurance metaphors, the frame of controllability, and buying time, while opponents draw on themes of uncertainty, governance, and the need to curb consumption wherein "problems of emission mitigation" are connected to "overeating and liposuction" (Luokkanen et al. 2013, 11). In a comprehensive study of media coverage of geoengineering from 2005 to 2013, Jonas Anshelm and Anders Hannson, in their article "The Last Change to Save the Planet? An Analysis of Geoengineering Advocacy Discourse in the Public Debate," assert that there are four major ways in which support for geoengineering has been framed. These include a discourse of "double fear," which expresses the danger and dilemmas of climate engineering research, the theme of intractability as it relates to addressing the problem of climate change politically, a framing of technological solutions as inevitable, and the claim that geoengineering is simply an extension of natural processes. The authors suggest that understanding these media frames reveals "an important divide in the public debate on geoengineering and nature" (Anshelm and Hansson 2014a, b, 117).

Governance

How to govern geoengineering, were it to be actively tested, researched and deployed, poses an important set of problems and obstacles. This is particularly the case since climate engineering techniques have global ramifications including the noted possible impacts on agriculture and food security, air pollution (sulfates), ecosystem effects, biodiversity and weather patterns to name a few. Geoengineering is also not monolithic, which means that CDR and SRM technologies would likely require different governance regimes. It might also be necessary to distinguish between the various phases of research, technology development, testing and deployment when considering possible governance models.

There are a host of governance proposals and recommendations that have been articulated by public and private parties with respect to how a regime of geoengineering regulation might be established. According to Bidisha Banerjee, these approaches tend to fall within two categories: The first is based on internationalist approach using existing treaties like the Convention on Biological Diversity (CBD), the United Nations Framework Convention on Climate Change (UNFCC), the 1972 London Convention, and 1996 London Protocol on the Prevention of Marine Pollution by Dumping of Wastes and Other Matter. The Convention on Biological Diversity has a moratorium on geoengineering in the form of ocean fertilization, but exempts 'legitimate research' broadly defined.

The second governance model consists of the adoption of a voluntary code of conduct set out by scientists working on climate engineering and climate change research (Banerjee 2011, 11). A recent example of this is can be found in a working paper put out by the Institute for Advanced Sustainability Studies (IASS) Potsdam titled 'Draft Code of Conduct for Responsible Scientific Research involving Geo-engineering.' In it, the authors call for the implementation of several protocols including: acting in accordance with environmental law; the assessment of proposals on an individual basis wherein States must examine the project's scientific attributes and environmental impacts; collaboration between States and international organizations; adoption of the precautionary principle; and a clear recognition of the need for post-project monitoring (Hubert and David 2015).

In addition to codes of conduct, a great deal of work has been done on formulating a set of governance principles that might guide further research. Jane C.S. Long and Dane Scott, in their article "Vested Interests and Geoengineering Research," call for traditional policy and research norms to be adhered to in any further research on geoengineering including transparency with respect to objectives and results, a public funding model, and the use of independent advisory committees with high levels of public consultation (Long and Scott 2013). These echo the Oxford Principles articulated below. Substantive public engagement with climate engineering in particular has also been thematized by university affiliated scientific bodies as well as government initiatives – both of which are also discussed further in the section on sulfate SRM.

The preceding sections have been intended to furnish the reader with some basic information about geoengineering as a scientific practice and technique with supplementary information about the key players (supporters and opponents), current status of research as well as issues surrounding governance and public participation. As stated, when taken as an ideal type, geoengineering can be understood as being constituted by technologies with general uniformities that lend themselves to the kind of analysis I perform in later Chapters. In the subsequent sections, I deliver a summary of stratospheric sulphate aerosol SRM geoengineering in preparation for its use as a case study in the Chapters that draw on FCE to generate novel insights and critiques of the SRM approach from a feminist perspective.

SRM

The case study, used throughout the Chapters, draw on feminist contextual empiricism (FCE) to examine climate engineering is stratospheric SRM geoengineering, which is a technique that aims to inject particles into the stratosphere that would scatter and reflect solar radiation away from the earth. Its objective is to increase the earth's albedo or reflectivity in a manner that mimics the cooling effects often observed after volcanic eruptions. One notable example, widely cited by proponents, is the 1991 Mount Pinatubo eruption in the Philippines, which is said to have reduced average temperature by approximately 0.5 K for about 2 years (Hansen et al. 1992; Soden et al. 2002; Minnis et al. 1993). In this section, I discuss current research on this technique ranging from contemporary scientific studies (mostly modeling), to proposed field-testing, as well more sociological and political analyses. Climate modeling poses a significant challenge and is discussed at length – particularly as it relates to later Chapters.

Current Research

Over the past few years, a number of studies have been conducted on sulphate geoengineering using complex climate modeling as well as historical evidence from volcanic eruptions. The purpose of these studies is to predict the ideal size and composition of particles, the most suitable place and time to inject particulates, the overall effects of this tactic, and its general effectiveness in reducing warming (Bluth et al. 1997; Farquhar and Roderick 2003; Ferraro et al. 2011). The size and composition of the particles is of particular importance in that they must remain consistently small enough to effectively reflect sunlight. Another possible limitation made manifest by these studies is the injection mechanism. Suggested options range from aircraft or artillery to tethered balloons or a large building.

A notable project that went quite far towards field-testing took place in the UK. The SPICE project, which stands for Stratospheric Particle Injection for

Climate Engineering, was aimed at investigating and testing the delivery mechanism of this approach in partnership between the Universities of Bristol, Cambridge, Edinburgh and Oxford (and supported financially by the EPSRC, NERC and STFC). It also intended to test ideal particles at a later date as well as to generate superior modeling practices (SPICE 2016). Investigators had intended to test the tethered balloon delivery method in isolation using water, but called off the experiment in 2012 due to a patent row, opposition from environmental groups, and concerns about governance. Significantly, a consultation process was conducted with members of the public prior to the tests' cancellation. Researchers at Bristol University, who conducted these sessions, found a general willingness to allow the test to occur, but also found significant discomfort with the stratospheric aerosols method (Pidgeon et al. 2013). Basic research on modeling and governance at the SPICE project is still ongoing.

With respect to ongoing research outside of North America and Western Europe, there are a several projects worth mentioning. Of particular note, in 2009, Russian scientist Yuri Izrael, with Russian government support, conducted a test very similar to the one proposed, but never executed, by the SPICE project and sprayed particulates "from a helicopter to assess how much sunlight was blocked by the aerosol plume" (Goldenberg et al. 2013). The result, as claimed by the team involved, indicated that the approach has the potential to mitigate global warming (Izrael et al. 2010). Critics, however, have challenged both the accuracy of test itself as well as the transparency of the assessment process (Melesheko et al. 2010). Recently, China has also accelerated their study of geoengineering with a number of in-depth projects that began in 2010 on solar radiation management. As Cao et al. outlines (2015), this includes detailed studies of the efficacy of different geoengineering schemes (Moore et al. 2010), an investigation into the effects of CO_2 and solar forcing on the climate (in terms of adjustments) (Cao et al. 2012), and a 2014 study of the effects of volcanic eruptions on monsoons in China (Zhuo et al. 2014).

Finally, in Finland, the Academy of Finland launched a 3 year program between 2011 and 2014 to, among other things, study the ability for commercial planes to deliver sulfur while transiting passengers as a mode of delivery for SRM climate engineering (Academy of Finland 2014). This was done through extensive modeling without plans for field tests.

Effectiveness and Risk

Generally, advocates of sulphate particle injection argued that, its cost are relatively low, based on estimates that include technical apparatus, the aerosols themselves etc. (McClellan et al. put it at 8BN a year at the high end (Brahic 2009)); its impacts are global, thereby increasing the likelihood it can change temperature globally – with estimates of radiative forcing at 5–10% if placed at an optimal place with ideal particulates (Wigley 2006); and it is technically possible to execute (Rasche et al. 2008).

Some advocates have also argued that because this technique extends natural processes that have been observed and studied at length (e.g. the effects of volcanic eruptions), it is less likely to have unintended consequences (Bates et al. 1992). It is also worth noting that the Royal Society Report (2009), in their assessment of individual geoengineering schemes, rated stratospheric aerosol geoengineering the highest with regard to the criteria of effectiveness, affordability and timeliness. A low score, however, was given on the standard of safety.

Yet, there are a whole host of possible side effects of the stratospheric injections approach that are troubling to say the least. The first is has to do with the likely effect on the ozone layer, which sulphate aerosols in particular might deplete by increasing the area on which chemical reactions that eat away at the ozone can take place (Nowack et al. 2016). Second, is the possible disruptive effect on levels of precipitation. It has been demonstrated that volcanic eruptions, like that of Pinatubo, led to drought in regions that rely on monsoons (Haywood et al. 2013). As Alan Robock argues, the impact this would have on agriculture in Asia and Africa could be catastrophic. Yet he notes that its severity would depend on other factors including "how much CO_2 fertilization…would compensate for the negative impacts of geoengineering, and how humans would adapt to the changing climate" (Robock 2014a, b, 176).

Another significant possible side effect is SRM's impact on the sky – which would no longer be blue. There are significant psychological and cultural impacts of a transformed horizon on the public at large associated with this that require attention. Another related consequence is that the solar energy supply would decrease and lead to a shortfall in a vital alternative to fossil fuels (Murphy 2009). Finally, abrupt cessation of sulphuric SRM poses a challenge with recent research suggesting that were large-scale sulphate seeding stopped, dangerous levels of rapid warming would likely follow (Ross and Matthews 2009; Brovkin et al. 2009). Yet, despite these negative consequences, there are possibilities of positive ones. For instance, with an increase of diffuse light, photosynthesizing plants might grow more effectively, thereby producing climate mitigating network effects. However, as Robock notes,

> While an increased carbon sink would be a benefit of stratospheric geoengineering, the effect would be felt differentially between different plant species, and whether it would help or hurt the natural ecosystem, or whether it would preferentially favour weeds rather than agricultural crops, has not been studied in detail yet (Robock 2014a, b, 177).

The Royal Society (2009) and National Academy of Sciences (2015a, b), in their appraisals of this climate engineering technique, articulate a number of barometers that reflect and encompass the possible side effects listed above, as well as the likelihood of success, on a more generalized level. They include: ethical acceptability; levels of confidence and risk; the need for robust regulatory frameworks; clarity on who gets to decide; what 'acceptable' risk entails; and an assessment of levels of responsibility. The NAS report highlights the fact that this approach will likely lead to ocean acidification and significant termination problems, as well as not addressing the underlying causes of excess CO_2 accumulation in the atmosphere.

Governance

With respect to governance, there are a variety of reports and studies that have engaged with this issue of sulfate geoengineering and regulation of which I discuss a few of the more noteworthy. What is interesting is that most articulate general principles and recommendations for this form of SRM, rather than substantive policies. The Oxford Principles are probably the most prominent example. They were submitted to and adopted by the UK House of Commons Science and Technology Select Committee's study of geoengineering in 2009. The substance of these principles is explained more fully below (Oxford Geoengineering Programme 2009).

The Royal Society, in 2011, launched an SRM governance initiative, SRMGI, in conjunction with the Academy of Sciences for the Developing World and the Environmental Defense Fund, which examines the various articulated principles on geoengineering governance from all over the world. Their study determined that more public participation in geoengineering research is needed and that traditional mitigation should still take the driver's seat. Moreover, they recommend that governance, in light of the different types of SRM, should be differentiated and only with a clearly defined regime in place should its use be contemplated.

Finally, the joint US (Congressional) and UK (Parliamentary) investigation into geoengineering concluded that the UNFCC, or an international consortium, should work to engage with the subject of geoengineering regulation as they are best placed to address concerns about both SRM and CDR research (House of Commons 2009–2010). They suggested a number of rules, based on the Oxford Principles, which recommends the regulation of geoengineering as a public good (with restrictions on funding and IPR applications); high levels of public participation; enhanced transparency with respect to results; independent assessment; avenues for compensation, and the inclusion of "legal, social, and ethical implications" which would provide "a sound foundation for developing future regulation" (House of Commons 2009–2010, 51).

Issues of Science, Modeling and Simulation

I have set aside the subjects of science, modeling and simulation until the end of this Chapter deliberately because basic knowledge of the terrain is needed. Here, I discuss modeling as it relates to geoengineering writ large as well as to sulphate SRM in particular. This section draws in insights from feminist approaches to science and technology, which provides a necessary framework from which to grapple with the themes of epistemological trust, risk, truth, accuracy, interpretation, and justification. While I do take up the perennial question of objective truth as it relates to scientific practice, my analysis in this book is limited to the process by which scientific consensus is reached and justification established, rather than on the more philosophical question of truth-content and world-truth relations.

Instead, my objective is to reconcile a pluralist, situated, and value laden science rooted in feminist science studies with a practice of science and scientific modeling that contains a large degree of uncertainty, and to articulate need for public and political confidence in climate science. In pursuit of this, I discuss what climate modeling is meant to do, how to deal with the uncertainties inherent in its findings; how feminist science intersects with these uncertainties in particular ways; and the productive conclusions that can be drawn from these intersections. This establishes a path through which to reconcile scientific uncertainty with epistemological plural-ism and, in doing so, takes care of a problematic line of argumentation which asserts that a pluralistic science endangers scientific consensus and that uncertainties in science cannot be productively reconciled. These are distinct claims which, if mis-understood, can have the effect of policy paralysis and distorted public perceptions of climate change. Having done this, the subsequent Chapters that draw on feminist science studies to critique geoengineering can be undertaken.

First, on the subject of uncertainty, the central question of note, regarding climate and geoengineering science and modeling, has to do with the status of the knowl-edge generated. First, how is it that we come to accept and act upon a set of scien-tific conclusions and modeling techniques that, by the admission of scientists themselves, are unquestionably complex, theoretically uncertain, and practically imprecise (Räisänen 2001; Brohan et al. 2006; Stern 2016)? And second, how does a pluralistic and feminist approach to science avoid the risk of lapsing into a 'any-thing goes' ontology and, instead, work to contribute productively towards climate action while simultaneously delivering a substantive critique of geoengineering.

Generally, climate models are meant to mathematically represent the interactions between the earth's atmosphere, oceans, ice, land and the sun as *trends* rather than singular events. They simplify real systems to aid with understanding how natural phenomenon, like oceans and the atmosphere, behave with the assistance of com-puters. These computers are needed to process the necessary data sets to simulate the climate. The mathematical and physical principles on which models are based are robust but, as representations, they are by definition imperfect. Currently, global climate models, also called global circulation models or GCMs, predict temperature well, but do not excel at predicting precipitation, cloud dynamics, flooding or hurricanes. They are also more effective at predicting continental and global rather than regional future climates (Bony and Dufresne 2005; Webb et al. 2013).

The initial conditions, parameters, and boundary conditions selected; the choice of spatial and temporal units; and uncertainties around human behavior adds to this complexity – particularly in light of the fact that it is individual scientists who make these value judgments. As Knutti et al. argues, boundary conditions, like the output of the sun, CO_2 concentrations, the position of ice-sheets, the precise composition of the atmosphere, and time-scales are "prescribed externally to the model, and results are often interpreted simply as conditional on the boundary conditions." Running and averaging multiple scenarios can mitigate this "but in most of the cli-mate change projections, uncertainties in the model structure and parameters domi-nate" (Knutti 2008, 4649; Oreskes et al. 1994; Stainforth et al. 2007).

Furthermore, this uncertainty has actually increased in light of the move from general circulation models (GCMs) to global climate models, which, according to Stevens and Bony, adds additional, sometimes burdensome, complexity to climate simulations (Steven and Bony 2014). Yet it is also the case that, over the past few years, significant improvements have been made to both modeling and simulation including better testing protocols; enhanced understanding of inter-model differences and feedbacks; better computing models and simulation techniques; the elimination of flux adjustments (which are thought to introduce biases); improved understanding of extreme weather events; and enhanced understanding of climate processes like El Nino-Southern Oscillation phenomenon (Randall et al. 2007, 591; Yukimoto and Noda 2002; Alcamo 2012). Also of note are rising levels of international cooperation through, for instance, the creation of cooperative model intercomparison projects like the Working Group on Coupled Models (WGCM), and their Coupled Model Intercomparison Project (CMIP) (Meehl et al. 2007; Giorgetta et al. 2013).

Modeling of geoengineering scenarios incorporate and extend these uncertainties because they require the same setting of parameters, boundaries, and initial conditions. Ambiguities associated with SRM modeling, for example, include the lack of a consistently clear representation of the impacts on local populations (Xia 2014), as well as retaining gaps with respect to how the carbon cycle will react to SRM (Kravitz et al. 2015). There are also high concerns regarding changes in precipitation patterns as well as "Missing external forcings (e.g. by absorbing aerosols) or errors in observations [which] could contribute to the discrepancy between observations and model simulations" (Hergerl, Solomon 2009) (Wentz et al. 2007; Lambert et al. 2008). Sulphate modeling also requires more expensive 3D models in order to more accurately represent regional specificities as well as the "heterogeneities in the resulting aerosol distribution" (Weisenstein et al. 2015, 11852), which limits the number to times it can be run.

In order to mitigate these uncertainties, as noted, intercomparison projects have also been launched such as the Geoengineering Model Intercomparison Project (GeoMIP) which was established in order to create a multimodel ensemble (MME) of several GCMs, as well as single model ensembles – all in order to better understand climate model responses to SRM geoengineering (Kravitz et al. 2011). Largely voluntary cooperation amongst modeling groups has greatly enhanced understanding of "certain methods of geoengineering, as well as the fundamental underpinnings of climate system response to perturbations" (Robock and Kravitz 2013). It is also the case that, for aerosols, findings can be compared to natural analogues – e.g. the Mount Pinatubo cooling – as well as being run as field tests if permitted.

Dealing with these complexities in ways that scientific uncertainty is not taken as an excuse for irreconcilable relativism and/or climate change denialism, while also allowing for the clear communication of recent improvements in technique and process, can be done in two ways: by policy or epistemology. I discuss each of these in turn. With respect to epistemology, this is where feminist science has much to offer. I take up this thesis quite briefly here, as it will be elaborated on in much more extensively in subsequent Chapters. Once more, the purpose of this explication is to deal with the issue of uncertainty such that the feminist critique of geoengineering that follows, of which values, background assumptions, and scientific pluralism are embraced and not rejected, is not accused of haphazard relativism. In this way, feminist science can stand on epistemologically firmer ground.

First, on the level of policy, concerns have been articulated with regard to how to explain acceptable levels scientific uncertainty to policy makers who not only need to formulate public policy in the face of this uncertainty, but also need to communicate and justify their decisions to an increasingly skeptical public. Over the past few years, this concern has been vocalized in prominent fora. In a review of the 2010 IPCC report, the InterAcademy Panel, a global network of national science academies, took the bold step of articulating the need to find a way to reconcile scientific uncertainty with sufficient confidence in order to move forward and create policies that are both assured and supported. As the InterAcademy Council makes clear,

> The evolving nature of climate science, the long time scales involved, and the difficulties of predicting human impacts on and responses to climate change mean that many of the results presented in IPCC assessment reports have inherently uncertain components. To inform policy decisions properly, it is important for uncertainties to be characterized and communicated clearly and coherently (InterAcademy Council 2010).

There are several ways in which this can be performed –the first of which is to restructure the public's understanding of what models can and cannot do. Without getting into epistemological questions here, this involves educating the public that models are idealized representations of some part of the climate system which can serve interpretive and/or predictive functions. As representations, they can mediate between theory and real world data and serve multiple knowledge generating roles. However, in light of the urgency associated with climate change, predictive models have come to be especially valued. Their veracity, as such, is of particular concern. If one of the goals of climate science is to provide decision-relevant information, as Knutti argues, this requires a delicate balancing act between "giving the most detailed information possible to guide policy versus communicating only what is known with high confidence" (Knutti 2008, 4060).

Another way in which uncertainty can be better managed is to conceptualize it as existing on a spectrum. Smith and Stern (2011) argued that, there are four types of non-mutually exclusive scientific uncertainty that can provide a basis from which it can be managed. First is imprecision or statistical uncertainty which covers outcomes that may not be precise, but which are robust and can form the basis on which decisions are made. Climate science fits into this category in that it can provide a reliable gage of climate change.

Second, is ambiguity, or recognized ignorance, which is an example of a second order uncertainty on which probabilities cannot be made. It is likely that geoengineering fits into this category. Third is intractability, wherein computations lie beyond current capacities – either mathematical or computational. Finally, there is indeterminacy, which is "relevant to policy-making for which no precise value exists. This applies, for instance, with respect to a model parameter that does not correspond to an actual physical quantity. It can also arise from the honest diversity of views among people, regarding the desirability of obtaining or avoiding a given outcome" (Smith and Stern 2011). Treating uncertainty in these ways can go far in mitigating policy paralysis and anti-science denialism.

For feminist contextual empiricism, uncertainty in scientific practice, including modeling, is seen as productive when understood as encouraging pluralism and multiplicity and working to expose the oftentimes distorting values and assumptions of idealized positivist science and the institutions that produce it. This does not mean, however, that we cannot trust scientific consensus or that there is no 'truth' but, rather, regarding both climate change and geoengineering, that there are other questions that must be asked, other models that can be used, novel controversies to be exposed, and alternative knowledge systems worth considering (Harding 2004a, b, 2008; Haraway 2008). FCE in particular is clear in its support for the empirical basis of scientific justification by relying on "mainstream scientific practices of systematic experimentation, observation, and recording," (Allen and Baber 1992, 3), yet seeking to acknowledge context and values (Hawkesworth 1989). Pluralism can thus be seen as encompassing the possibility of alternative theories and models emanating from local situations where evidential standards are met. This line of thinking will be returned to further on.

Conclusion

The objective of this overview has been to provide those interested with basic knowledge of climate engineering as well as some deeper explication of the sulphate aerosol SRM technique that constitutes the case study. While gender does not figure as prominently here as in the Chapters that follow, what has been covered is crucial..

It is significant that is that in regards to geoengineering, the questions asked, assumptions made, research areas focused on, and the models themselves tend to be conventional or traditional in the sense that they are based on scientific practices that, in accordance with Enlightenment thought, assumes the existence of an objective, immutable external world that can be known through the application of empirical methods and reason (i.e. the scientific method). Feminist critiques of this approach are widespread (Haraway 1991; Best 1991; Harding 1986a, b) and will be taken up in subsequent Chapters – particularly with respect to FST and alternative epistemologies. Yet, importantly, it will do so without drifting into relativism

The concluding Chapter which takes up feminist technoscience (or technofeminism) focuses specifically on how the technological applications of the geoengineering itself are gendered and how its symbolic representation, in media and computer modeling, are as well. In each of the following Chapters, geoengineering is treated as a set of techniques with the common characteristics of global consequence, invasiveness (into the natural world), technological dependence (as a potential quick fix), and potential risk.

All other Chapters are dedicated to discuss Helen Longino's scientific virtues, as articulated in FCE, to assess how and in what ways geoengineering fails to meet the criteria of empirical adequacy, novelty, ontological heterogeneity, complexity or mutuality of interaction, applicability to human needs and decentralization of power or universal empowerment (Longino 1990). Longino's approach feminist science is based on in the practice of what she calls 'doing science as a feminist' which aims to expand human potentiality and challenge both traditional sex roles and "institutions of domination" such that we begin to insist on the "capacity of humans—male and female—to act on perceptions of self and society and to act to bring about changes in self and society on the basis of those perceptions" (Longino 2005, 205).

References

Abate, R. S. (2013). Ocean iron fertization: Science, law and uncertainty. In W. C. G. Burns & A. L. Strauss (Eds.), *Climate change geoengineering: Philosophical perspectives, legal issues, and governance frameworks*. Cambridge: Cambridge University Press.

ABN, Biofuelwatch, Gaia Foundation. (2009). *Biochar land grabbing: The impacts on Africa.* ABN, Biofuelwatch, Gaia Foundation. http://www.biofuelwatch.org.uk/docs/biochar_africa_briefing.pdf. Accessed 29 Oct 2016.

Academy of Finland. (2014). *Researchers look into geoengineering possibilities.* Academy of Finland http://web.archive.org/web/20150314161010/http://www.aka.fi/en-GB/A/Programmes-and-cooperation/Academy-programmes/Etusivun-elementit/Researchers-look-into-geoengineering-possibilities/. Accessed 13 Oct 2016.

Alcamo, J. (Ed.). (2012). *IMAGE 2.0: Integrated modeling of global climate change.* New York: Springer Science & Business Media Press.

Allen, K. R., & Barber, K. M. (1992). Ethical and epistemological tensions in applying a postmodern perspective to feminist research. *Psychology of Women Quarterly, 16*, 1–15.

Amelung, D., & Funke, J. (2014). Laypeople's risky decisions in the climate change context: Climate engineering as a risk-defusing strategy? *Human and Ecological Risk Assessment, 21*, 533–559.

American Meteorological Society. (2009). Geoengineering the climate system: A policy statement of the American Meteorological Society. *Bulletin of the American Meteorological Society, 90*, 1369–1370.

Ammann, C. M., Washington, W. M., Meehl, G. A., Buja, L., & Teng, H. Y. (2010). Climate engineering through artificial enhancement of natural forcings: Magnitudes and implied consequences. *Journal of Geophysical Research-Atmospheres, 115*, D22109.

Anderegg, W. R. L. (2010, June 21) Expert credibility in climate change. *Proceedings of the National Academy of Sciences, 107*(27), 12107–12109.

Anshelm, J., & Hansson, A. (2014a). The last chance to save the planet? An analysis of the geoengineering advocacy discourse in the public sphere. *Environmental Humanities, 3*, 101–123.

Anshelm, J., & Hansson, A. (2014b). Battling Promethean dreams and Trojan horses: Revealing the critical discourses of geoengineering. *Energy Research & Social Science, 2*, 135–144.

Asilomar International Conference on Climate Intervention Technologies. (2010, March). Asilomar Conference Center, California. https://www.scribd.com/document/25251727/The-Asilomar-International-Conference-On-Climate-Intervention-Geoengineering-Technologies-2010

Balint, P., et al. (2011). *Wicked environmental problems: Managing uncertainty and conflict.* Washington, DC: Island Press.

Banerjee, B. (2011). The limitations of geoengineering governance in a world of uncertainty. *Stanford Journal of Law, Science & Policy, 4*, 15–36.

Bates, S. S., Lamb, B. K., Guenther, A., Dignon, J., & Stoiber, R. E. (1992). Sulfur emissions to the atmosphere from natural sources. *Journal of Atmospheric Chemistry., 14*, 315–337. https://doi.org/10.1007/BF00115242.

Bellamy, R., Chilvers, J., Vaughan, N. E., & Lenton, T. M. (2012). A review of climate geoengineering appraisals. *Wiley Interdisciplinary Reviews-Climate Change, 3*(6), 597–615. https://doi.org/10.1002/wcc.197.

Best, S. (1991). Chaos and entropy: Metaphors in postmodern science and social theory. *Science as Culture, 2*(11), 188–226.

Blackstock, J. (2012). Researchers can't regulate climate engineering alone. *Nature, 486*(7402), 159–159. https://doi.org/10.1038/486159a.

Blain, S., Sarthou, G., & Laan, P. (2008). Distribution of dissolved iron during the natural iron-fertilization experiment KEOPS (Kerguelen Plateau, Southern Ocean). *Deep-Sea Research Part Ii-Topical Studies in Oceanography, 55*(5-7), 594–605.

Bluth, G. J. S., Rose, W. I., Sprod, I. E., & Krueger, A. J. (1997). Stratospheric loading of sulfur from explosive volcanic eruptions. *Journal of Geology, 105*, 671–683.

Bonney, R. (1996). Citizen science: A lab tradition. *Living Bird, 15*(4), 7–15.

Bony, S., & Dufresne, J. F. (2005). Marine boundary layer clouds at the heart of tropical cloud feedback uncertainties in climate models. *Geophysical Research Letters, 32*, 20. https://doi.org/10.1029/2005GL023851. Accessed 15 Oct 2016.

Bowker, G. C. (1992). What's in a patent? In W. E. Bijker & J. Law (Eds.), *Shaping technology/building society: Studies in sociotechnical change* (pp. 53–74). London: MIT Press.

Brahic, C. (2009, February 25). *Hacking the planet: The only climate solution left?* Reed Business Information Ltd. Accessed 28 Sept 2016.

Branson, R. (2016). *Virgin Earth challenge.* http://www.virginearth.com. Accessed 8 Oct 2016.

Brohan, P., et al. (2006). Uncertainty estimates in regional and global observed temperature changes: A new dataset from 1850. *Journal of Geophysical Research, 111*, D12106. https://doi.org/10.1029/2005JD006548. Accessed 5 Oct 2016.

Brovkin, V., Petoukhov, V., Claussen, M., Bauer, E., Archer, D., & Jaeger, C. (2009). Geoengineering climate by stratospheric sulphur injections: Earth system vulnerability to technological failure. *Climate Change, 92*, 243–259.

Burns, W. C. G., & Flegel, J. A. (2015). Climate geoengineering and the role of public deliberation: A comment on the US National Academy of Sciences' recommendations on public participation. *Climate Law, 5*, 252–294.

Burns, W. C. G., & Strauss, A. L. (2013). *Climate change geoengineering: Philosophical perspectives, legal issues and governance frameworks.* Cambridge: Cambridge University Press.

Cairns, R. C. (2014). Climate geoengineering: Issues of path-dependence and socio-technical lock-in. *Wiley Interdisciplinary Reviews-Climate Change, 5*(5), 649–661. https://doi.org/10.1002/wcc.296.

Cao, L., Bala, G., & Caldeira, K. (2012). Climate response to changes in atmospheric carbon dioxide and solar irradiance on the time scale of days to weeks. *Environmental Research Letters, 7*, 034015.

Cao, L., Gao, C. C., & Zhao, L. Y. (2015). Geoengineering: Basic science and ongoing research efforts in China. *Advances in Climate Change Research, 6*(3-4), 188–196.

Connick, S., & Judith Innes, J. (2003). Outcomes of collaborative water policy making: Applying complexity thinking to evaluation. *Journal of Environmental Planning and Management, 46*, 177–197.

Cook, J., et al. (2016, April 13). Consensus on consensus: A synthesis of consensus estimates on human-caused global warming. *Environmental Research Letters, 11*(4). https://doi.org/10.1088/1748-9326/11/4/048002.

Crutzen, P. (2006). Albedo enhancement by stratospheric sulfur injections: A contribution to resolve a policy dilemma? *Climate Change, 77*, 211–219.

Dean, J. (2009). Iron fertilization: A scientific review with international policy recommendations. *Environs: Environmental Law and Policy, 32*, 321–329.

Delanty, G. (1997). *Social science: Beyond constructivism and realism.* Minneapolis: University of Minnesota Press.

Doran, P. T., & Zimmerman, M. K. (2009). Examining the scientific consensus on climate change. *Eos Transactions American Geophysical Union, 90,* 3. https://doi.org/10.1029/2009EO030002.

Dunlap, R. E., & Jacques, P. J. (2013). Climate change denial books and conservative think tanks: Exploring the connection. *American Behavioral Scientist.* https://doi.org/10.1177/0002764213477096. Accessed 21 Oct 2016.

ETC Group. (2015). *Climate & geoengineering.* ETC Group. http://www.etcgroup.org/issues/climate-geoengineering. Accessed 12 Oct 2016.

Farquhar, G. D., & Roderick, M. L. (2003). Pinatubo, diffuse light, and the carbon cycle. *Science, 299,* 1997–1998. https://doi.org/10.1126/science.1080681.

Feenberg, A. (1999). *Questioning technology.* London/New York: Routledge.

Ferraro, A. J., Highwood, E. J., & Charlton-Perez, A. J. (2011). Stratospheric heating by potential geoengineering aerosols. *Geophysical Research Letters, 38,* L24706. https://doi.org/10.1029/2011GL049761.

Fleming, J. R. (2006). The pathological history of weather and climate modification: Three cycles of promise and hype. *Historical Studies of Physical and Biological Sciences, 37*(1), 3–25.

Fleming, J. R. (2007). The climate engineers. *The Wilson Quarterly, 31*(2), 46–60.

Fortmann, L., Ballard, H., & Sperling, L. (2008). Change around the edges: Gender analysis, feminist methods, and sciences of terrestrial environments. In L. Schiebinger (Ed.), *Gendered innovations in science and engineering* (pp. 79–96). Stanford: Stanford University Press.

Gardiner, S. M. (2010). Is arming the future with geoengineering really the lesser Evil? Some doubts about the ethics of intentionally manipulating the climate system. In S. M. Gardiner, S. Caney, D. Jamieson, & H. Shue (Eds.), *Climate ethics: Essential readings* (pp. 284–312). New York: Oxford University Press.

Geoenginering 2012. (2012). *Geo-engineering in the context of sustainable Part 2: Geo-engineering.* Geoengineering 2012. Available at: https://www.ecologyandsociety.org/vol17/iss1/art24/#experiments. Accessed 5 May 2018.

Gibbons, M. (1999). Science's new social contract with society. *Nature, 402,* C81–C84.

Giorgetta, M. A., et al. (2013). Climate and carbon cycle changes from 1850 to 2100 in MPI-ESM simulations for the Coupled Model Intercomparison Project phase 5. *Journal of Advances in Modeling Earth Systems, 5*(3), 572–597.

Goldenberg, S., Lukaks, M., & Vaughan, A. (2013, September 19). Russia urges UN climate report to include geoengineering. *The Guardian.* https://www.theguardian.com/environment/2013/sep/19/russia-un-climate-report-geoengineering. Accessed 6 Oct 2016.

Government Accountability Office (GAO). (2010). *Climate engineering: Technical status, future directions and potential responses.* GAO. http://www.gao.gov/highlights/d11711high.pdf. Accessed 4 Oct 2016.

Hamilton, C. (2011). *Ethical anxieties about geoengineering: Moral hazard, slippery slope and playing God.* Conference of the Australian Academy of Science. http://www.climate-engineering.eu/single/hamilton-clive-conference-paper-ethical-anxieties-about-geoengineering.html. Accessed 29 Sept 2016.

Hamilton, C. (2013). *How Bill Gates is engineering the Earth to resist climate change.* Crikey, 26 Feb 2013. https://www.crikey.com.au/2013/02/26/how-bill-gates-is-engineering-the-earth-to-resist-climate-change/. Accessed 23 Feb 2017.

Hansen, J., Lacis, A., Ruedy, R., & Sato, M. (1992). Potential climate impact of Mount Pinatubo eruption. *Geophysical Research Letters, 19,* 215–218.

Hansson, S. O. (2008). From the casino to the jungle. *Synthese, 168,* 423–432.

Haraway, D. J. (1991). *Simians, cyborgs, and women: The reinvention of nature.* New York: Routledge.

Haraway, D. (2008). Otherworldly conversations, terran topics, local terms. *Material Feminisms, 3,* 157.

Harding, S. (1986a). *The science question in feminism.* Ithaca: Cornell University Press.

Harding, S. (1986b). The instability of the analytical categories of feminist theory. *Signs: Journal of Women in Culture and Society, 11*(4), 645–664.

Harding, S. (1991). *Whose science? Whose knowledge? Thinking from women's lives*. Ithaca: Cornell University Press.

Harding, S. (2004a). A socially relevant philosophy of science? Resources from standpoint theory's controversiality. *Hypatia, 19*, 25–47.

Harding, S. G. (2004b). *The feminist standpoint theory reader: Intellectual and political controversies*. Hove: Psychology Press.

Harding, S. (2008). *Sciences from below: Feminisms, postcolonalities, and modernities*. Durham: Duke University Press.

Harper, K. C. (2008). Climate control: United States weather modification in the cold war and beyond. *Endeavour, 32*(1), 20–26.

Hawkesworth, M. E. (1989). Knowers, knowing, known: Feminist theory and claims of truth. *Signs, 14*, 533–557.

Haywood, J. M., Jones, A., Bellouin, N., & Stephenson, D. (2013). Asymmetric forcing from stratospheric aerosols impacts Sahelian rainfall. *Nature Climate Change, 3*, 660. https://doi.org/10.1038/nclimate1857.

Healey, P. (2014). *The stabilisation of geoengineering: Stabilising the inherently unstable*. Climate Geoengineering Governance Working Paper Series.

Heibel, M. (2013). Terra furura 2013: Interview with Vandana Shiva about geoengineering. *NoGeoingegneria*.http://www.nogeoingegneria.com/interviste/terra-futura-2013-interview-with-vandana-shiva-about-geoengineering/. Accessed 3 Oct 2016.

Hergerl, B., & Solomon, S. (2009). Risks of climate engineering. *Science Express*. https://doi.org/10.1126/science.1178530. Accessed 1 Nov 2016.

House of Commons. (2009–2010). *The regulation of geoengineering: Fifth report of session*.http://www.publications.parliament.uk/pa/cm200910/cmselect/cmsctech/221/221.pdf. Accessed 30 Sept 2016.

Hubert, A. M., & David, R. (2015). *Research involving geoengineering: Introduction, draft articles and commentaries*. Institute for Advanced Sustainability Studies. http://www.insis.ox.ac.uk/fileadmin/images/misc/An_Exploration_of_a_Code_of_Conduct.pdf. Accessed 30 Sept 2015.

Incorpera, F. P. (2016). *Climate change: A wicked problem. Complexity and uncertainty at the intersection of science, economics, politics and human behavior*. Cambridge: Cambridge University Press.

InterAcademy Council. (2010). *Climate change assessments: Review of the processes and procedures of the IPCC*.http://reviewipcc.interacademycouncil.net. Accessed 4 Oct 2016.

IPCC. (2013). *Climate change 2013: The physical science basis*. http://www.ipcc.ch/report/ar5/wg1/. Accessed 14 Oct 2016.

Ipsos-MORI. (2010). *Experiment Earth: Report on a public dialogue on geoengineering*. Natural Environment Research Council, Swindon UK. See http://www.nerc.ac.uk/about/consult/geo-engineering-dialogue-final-report.pdf. Accessed 1 Oct 2016.

Izrael, Y. A., et al. (2010). Field studies of a geo-engineering method of maintaining a modern climate with aerosol particles. *Russian Meteorology and Hydrology, 34*(10), 635–638.

Jones, A., Haywood, J., & Boucher, O. (2009). Climate impacts of geoengineering marine stratocumulus clouds. *Atmospheres: Journal of Geophysical Research, 114*, 9. https://doi.org/10.1029/2008JD011450.

Kalidindi, S., et al. (2015). Modeling of solar radiation management: A comparison of simulations using reduced solar constant and stratospheric sulphate aerosols. *Climate Dynamics, 44*(9–10), 2909–2925.

Keith, D. W. (2000). Geoengineering the climate: History and prospect. *Annual Review of Energy and the Environment, 25*(1), 245–284.

Keith, D. (2015). A cheap but dangerous global warming fix. *PBS Newshour*.http://www.pbs.org/newshour/making-sense/cheap-controversial-solution-climate-change/. 15 Oct 2015.

Kintisch, E. (2012). Overview of climate engineering. *The Bridge on Frontiers of Engineering, 42*(4), 5–9.

Knutti, R. (2008). Should we believe model predictions for future climate change? *Philosophical Transactions of the Royal Society A, 366*, 4647–4664.

Kravitz, B., et al. (2011). The geoengineering model intercomparison project (GeoMIP). *Atmospheric Science Letters, 12*, 162–167.

Kravitz, et al. (2015). The geoengineering model intercomparison project phase 6 (GeoMIP6): Simulation design and preliminary results. *Journal of Geoscientific Model Development, 3*, 3379–3392.

Lambert, F. H., et al. (2008). How much will precipitation increase with global warming? *Eos, 89*, 193–194.

Long, J. C. S., & Scott, D. (2013, Spring). Vested interests and geoengineering research. *Issues.* http://issues.org/29-3/long-4/. Accessed 22 Sept 2016.

Longino, H. E. (1990). *Science as social knowledge: Values and objectivity in scientific inquiry.* Princeton: Princeton University Press.

Longino, H. E. (2005). Can there be a feminist science? In A. E. Cudd & R. O. Andreasen (Eds.), *Feminist theory: A philosophical anthology* (pp. 210–217). Oxford/Malden: Blackwell Publishing.

Lou, X., et al. (2012). Cloud-resolving model for weather modification in China. *Chinese Science Bulletin, 57*, 1055–1061.

Lucaks, M. (2012, October 15). World's biggest geoengineering experiment 'violates' UN rules. *The Guardian.* https://www.theguardian.com/environment/2012/oct/15/pacific-iron-fertilisation-geoengineering. Accessed 5 Oct 2016.

Lunt, D. J. (2013). Sunshades for solar radiation management. In T. Lenton & N. Vaughan (Eds.), *Geoengineering responses to climate change* (pp. 9–20). New York: Springer.

Luokkanen, M., Huttunen, S., & Hilden, M. (2013). Geoengineering, news media and metaphors: Framing the controversial. *Public Understanding of Science*, 1–16. http://pus.sagepub.com/content/early/2013/02/14/0963662513475966. Accessed 4 Oct 2016.

Marshall, G. (2014). *Don't even think about it: Why our Brains are wired to ignore climate change.* New York: Bloomsbury Publishing.

Mathews, H. D., & Caldeira, K. (2007). Transient climate-carbon simulations of planetary geo-engineering. *Proceedings of the National Academy of Sciences, 104*, 9949–9954. https://doi.org/10.1073/pnas.0700419104.

Matthews, H. D., & Caldeira, K. (2007). Transient climate–carbon simulations of planetary geoengineering. *Proceedings of the National Academy of Sciences, 104*(24), 9949–9954.

McClellan, J., Sisco, J., Suarez, B., & Keogh, G. (2010). *Geoengineering cost analysis: Final report.* Cambridge MA: Aurora Flight Sciences Corporation.

Meehl, G. A., et al. (2007). The WCRP CMIP3 multi-model dataset: A new era in climate change research. *Bulletin of the American Meteorological Society, 88*, 1383–1394.

Meleshko, V. P., Kattsov, V. M., & Karol, I. L. (2010). Is aerosol scattering in the stratosphere a safety technology preventing global warming? *Russian Meteorology and Hydrology, 35*(7), 433–440.

Mercer, A. M., Keith, D. W., & Sharp, J. D. (2011). Public understanding of solar radiation management. *Environmental Research Letters, 6*, 044006.

Minnis, P., et al. (1993). Radiative climate forcing by the Mount Pinatubo eruption. *Science, 259*, 1411–1415.

Moore, J., Jevrejeva, S., & Grinsted, A. (2010). Efficacy of geoengineering to limit 21st century sea-level rise. *Proceedings of the National Academy of Sciences U S A, 107*, 15699–15703.

Murphy, D. M. (2009). Effect of stratospheric aerosols on direct sunlight and implications for concentrating solar power. *Environment Science Technology, 43*(8), 2784–2786. https://doi.org/10.1021/es802206b.

NAS. (2015a). Climate intervention: Reflecting sunlight to cool the Earth. *NAS.* https://nas-sites.org/americasclimatechoices/other-reports-on-climate-change/climate-intervention-reports/. Accessed 30 Sept 2016

NAS. (2015b). Climate intervention: Carbon dioxide removal and reliable sequestration. *NAS*. https://nas-sites.org/americasclimatechoices/other-reports-on-climate-change/climate-intervention-reports/. Accessed 30 Sept 2016

National University of Singapore Blog. (2015). *Two harvard engineers and a balloon*. National University Singapore. Accessible at: http://blog.nus.edu.sg/pollutionsolution/. Accessed 2 May 2018.

Niemeier, U., Schmidt, H., Alterskjær, K., & Kristjánsson, J. E. (2013). Solar irradiance reduction via climate engineering: Impact of different techniques on the energy balance and the hydrological cycle. *Journal of Geophysical Research: Atmospheres, 118*(21), 11905–11917.

Nowack, P. J., et al. (2016). Stratospheric ozone changes under solar geoengineering: Implications for UV exposure and air quality. *Atmospheric Chemistry and Physics, 16*, 4191–4203.

Oldham, P., et al. (2014). Mapping the landscape of climate engineering. *Philosophical Transactions of the Royal Society A, 372*, 1–20. https://doi.org/10.1098/rsta.2014.0065.

Oreskes, et al. (1994). Verification, validation and confirmation of numerical models in the earth sciences. *Science, 263*, 641–646.

Oxford Geoengineering Programme. (2009). *The Oxford principles*.http://www.geoengineering.ox.ac.uk/oxford-principles/principles/. Accessed 15 Sept 2016.

Oxford Geoengineering Programme. (2016). *What is geoengineering?*http://www.geoengineering.ox.ac.uk/what-is-geoengineering/what-is-geoengineering/. Accessed 1 Sept 2016.

Parkhill, K., & Pidgeon, N. (2011). *Public engagement on geoengineering research: Preliminary report on the SPICE deliberative workshops*. Understanding Risk Working Paper 11-01. School of Psychology, Cardiff University UK.

Pidgeon, et al. (2012). Exploring early public responses to geoengineering. *Philosophical Transaction of the Royal Society, 370*, 1974. https://doi.org/10.1098/rsta.2012.0099. Accessed 1 Oct 2016.

Pidgeon, N., et al. (2013). Deliberating stratospheric aerosols for climate geoengineering and the SPICE project. *Nature Climate Change, 3*, 451–457.

Räisänen, J. (2001). CO_2-induced climate change in CMIP2 experiments: Quantification of agreement and role of internal variability. *Journal of Climate, 14*, 2088–2104.

Randall, D. A., et al. (2007). Climate models and their evaluation. In S. D. Solomon et al. (Eds.), *Climate change 2007: The physical science basis. Contribution of Working Group I to the Fourth Assessment Report of the IPCC (FAR)* (pp. 589–662). Cambridge: Cambridge University Press.

Rasch, P. J, Tilmes, S, Turco, R. P, Robock, A Oman, L, Chen, C, Stenchikov, G. L, & Garcia, R. R. (2008, November). An overview of geoengineering of climate using stratospheric sulphate aerosols. *Philosophical Transactions. Series A Mathematical, Physical, and Engineering Sciences, 366*(1882), 4007–4037.

Rayner, S., et al. (2013). The Oxford principles. *Climatic Change, 121*(3), 499–512.

Ricke, K., Morgan, G., & Allen, M. (2010). Regional climate response to solar-radiation management. *Nature Geoscience, 3*, 8.

Rittel, H. W. J. (1973). Dilemmas in a general theory of planning. *Policy Sciences, 4*, 155–169.

Robock, A. (2008a). 20 reasons why geoengineering may be a bad idea. *Bulletin of the Atomic Scientists, 64*(2), 14–18.

Robock, A. (2008b). Whither geoengineering? *SCIENCE-NEW YORK THEN WASHINGTON, 320*(5880), 1166.

Robock, A. (2014a). Geoengineering the climate system. In R. E. Hester & R. M. Harrison (Eds.), *Issues in environmental science and technology* (pp. 162–185). Cambridge: Royal Society of Chemistry.

Robock, A. (2014b). Stratospheric aerosol geoengineering. In R. E. Hester & R. M. Harrison (Eds.), *Geoengineering of the climate* (pp. 162–185). Cambridge: The Royal Society of Chemistry.

Robock, A., & Kravitz, B. (2013). *Use of models, analogs and field-tests for geoengineering research*. Opinion Article, Geoengineering Our Climate Working Paper and Opinion Article Series. http://wp.me/p2zsRk-99. Accessed 16 Oct 2016.

Robock, A., Oman, L., & Stenchikov, G. L. (2008). Regional climate responses to geoengineering with tropical and Arctic SO$_2$ injections. *Journal of Geophysical Research: Atmospheres, 113*, D1601–D16101. https://doi.org/10.1029/2008JD010050.

Ross, A., & Matthews, H. D. (2009). Climate engineering and the risk of rapid climate change. *Environmental Research Letters, 4*, 045103.

Rowe, G., & Frewer, L. J. (2000). Public participation methods: A framework for evaluation. *Science, Technology and Human Values, 25*, 3–29.

Sarewitz, D. (2010). Not by experts alone. *Nature, 466*, 688.

Schmitz, O. J. (2016a, January 25). How 'natural geoengineering can help stop global warming' *Yale Environment 360*. http://e360.yale.edu/feature/how_natural_geo-engineering_can_help_slow_global_warming/2951/. Accessed 24 Sept 2016.

Schmitz, O. J. (2016b, January 25). How 'natural geoengineering' can help slow global warming. *e360 Yale University.*http://e360.yale.edu/feature/how_natural_geo-engineering_can_help_slow_global_warming/2951/. Accessed 23 Sept 2016.

Seidenfeld, M. (2000). Empowering stakeholders: Limits on collaboration for flexible regulation. *William and Mary Law Review, 41*, 411.

Shepherd, J., et al. (2009). *Geoengineering the climate: Science, governance and uncertainty.* London: The Royal Society.

Smith, L. A., & Stern, N. (2011). Uncertainty in science and its role in climate policy. *Philosophical Transactions of the Royal Society A, 369*(1956), 4818–4841.

Smoker, R. (2013, March 22). *Geoengineering is a dangerous solution to climate change, Huffingtonpost.* http://www.huffingtonpost.ca/rachel-smolker/geoengineering-climate-change_b_2907068.html. Accessed 26 Sept 2016.

Soden, B. J., Wetherald, R. T., Stenchikov, G. T., & Robock, A. (2002). Global cooling after the eruption of Mount Pinatubo: A test of climate feedback by water vapor. *Science, 296*, 727–730.

Solomon, S., & Oceanic, N. (1999). Stratospheric ozone depletion: A review of concepts and history. *Review of Geophysics, 37*, 275–316.

Specter, M. (2012a, May 14). The climate fixers. Annals of Science, *The New Yorker.*http://www.newyorker.com/magazine/2012/05/14/the-climate-fixers. Accessed 14 Oct 2016.

Specter, M. (2012b, October 18). The first geo-vigilante. *The Nation.* http://www.newyorker.com/news/news-desk/the-first-geo-vigilante. Accessed 5 Oct 2016.

SPICE. (2016). *Aims and background.* http://www.spice.ac.uk/about-us/aims-and-background/. Accessed 19 Sept 2016.

Stainforth, D. A., et al. (2007). Confidence, uncertainty and decision-support relevance in climate predictions. *Philosophical Transactions of the Royal Society of London A: Mathematical, Physical and Engineering Sciences, 365*(1857), 2145–2161.

Stern, N. (2016). Economics: Current climate models are grossly misleading. *Nature, 530*(7591), 407–409.

Stevens, B., & Bony, S. (2014). What are climate models missing? *Science, 340*(6126), 1053–1054.

Stolaroff, J. K., Keith, D. W., & Lowry, G. V. (2008). Carbon dioxide capture from atmospheric air using sodium hydroxide spray. *Environmental Science & Technology, 42*(8), 2728–2735.

Taubenfeld, R. F., & Taubenfeld, J. (2009). Some international implications of weather modification activities. *International Organization, 23*(4), 808–833.

The Keith Group. (2016). *About us.* http://keith.seas.harvard.edu. Accessed 1 Sept 2016.

The Royal Society. (2009, September). *Geoengineering the climate: Science, governance and uncertainty.* London: Royal Society of London. https://royalsociety.org/topics-policy/publications/2009/geoengineering-climate/. Accessed 17 Sept 2016.

Vidal, J. (2012a). Bill Gates backs climate scientists lobbying for large-scale geoengineering. *The Guardian,* 6 February 2012. https://www.theguardian.com/environment/2012/feb/06/bill-gates-climate-scientists-geoengineering. Accessed 13 Feb 2017.

Vidal, J. (2012b, February 6) Bill Gates backs climate scientists lobbying for large-scale geoengineering. *The Guardian.*https://www.theguardian.com/environment/2012/feb/06/bill-gates-climate-scientists-geoengineering. Accessed 4 Oct 2012.

Webb, M. J., Lambert, F. H., & Gregory, J. M. (2013). Origins of differences in climate sensitivity, forcing and feedback in climate models. *Climate Dynamics, 40*, 677. https://doi.org/10.1007/s00382-012-1336-x. Accessed 13 Oct 2016.

Weisenstein, D. K., Keither, D. W., & Dykema, J. A. (2015). Solar engineering using solid aerosol in the stratosphere. *Atmospheric Chemistry and Physics, 15*, 11835–11859.

Wentz, F. J., Ricciardulli, L., Hilburn, K., & Mears, C. (2007). How much more rain will global warming bring? *Science, 317*, 233. https://doi.org/10.1126/science.1140746. Accessed 1 Nov 2016.

Wigley, T. M. L. (2006). A combined mitigation/geoengineering approach to climate stabilization. *Science, 314*(5798), 452–454.

Willoughby, H. E., Jorgensen, D. P., Black, A., & Rosenthal, S. L. (1985). *Project STORMFURY: A scientific chronicle 1962–1983*. Hurricane Research Division, AOML/NOAA. https://doi.org/10.1175/1520-0477(1985)066<0505:PSASC>2.0.CO;2. Accessed 25 Sept 2016.

Winner, L. (1980). Do artifacts have politics? *Daedalus, 109*, 121–136.

World Meteorological Organization (WMO). (2013). *6th Joint Science Committee of the World Weather Research Programme*. http://www.wmo.int/pages/prog/arep/wwrp/new/documents/Doc_3_6_weather_mod_2013_Final_tn.pdf. Accessed 3 Oct 2016.

Xia, L. (2014). Solar radiation management impacts on agriculture in China: A case study in the Geoengineering Model Intercomparison Project (GeoMIP). *Journal of Geophysical Research, 119*, 8695–8711.

Yukimoto, S., & Noda, A. (2002). Improvements of the meteorological research institute global ocean-atmosphere coupled GCM (MRI-CGCM2) and its climate sensitivity. *National Institute for Environmental Studies, Tsukuba, Japan, 37*, 44.

Zhang, Y., & Posch, A. (2014). The wickedness and complexity of decision making in geoengineering. *Challenges, 5*, 290–408.

Zhuo, Z., Gao, C., & Pan, Y. (2014). Proxy evidence for China's monsoon precipitation response to volcanic aerosols over the past seven centuries. *Journal of Geophysical Atmospheric Research, 119*, 6638–6652.

Chapter 3
FCE and Empirical Adequacy

Abstract This chapter provides an overview of Helen Longino's approach to feminist science studies beginning with a discussion of her stand on objectivity, pluralism, sociality, values, and feminist scientific virtues. An overview of each of the virtues of empirical adequacy, novelty, heterogeneity, mutuality or reciprocity of interaction, applicability to human needs and diffusion of power or universal empowerment is also provided. This sets the stage for the analysis of solar climate engineering in relation to her feminist virtues.

Keywords Empirical adequacy · Novelty · Heterogeneity · Interactionism · Human needs · Power · Helen Longino · Geoengineering · Constituitive · Underdetermination · Pragmatic.

Geoengineering lends itself to a feminist critique from a variety of perspectives – particularly when understood as a set of scientific and socio-cultural techniques with the potential of manifesting itself in the form of powerful technological artifacts. There has been very little work done in this area apart from a 2014 article in *Hypatia* titled, "Gender and Geoengineering." In it, the authors unpack the geoengineering/gender nexus beginning with the actors involved in endorsing geoengineering who are, unsurprisingly, mostly men. In the piece, Buck et al. point out that this not only reflects imbalances in decision-making power, but also affects how the geoengineering gets framed in the media. Ecofeminist concerns regarding the implications of human intervention in the environment and the continued use of a reductionist approach to nature is also critiqued as failing to put "the lives of the most vulnerable…at the center of ethical concern" (Buck et al. 2014, 661). A nonreductionist, inclusive and contextual conception of science and technology is posited as a potential alternative, particularly as it relates to the future development of geoengineering – a subject on which they remain agnostic.

Building on this, throughout the following Chapters discussions of representation in both science and media are discussed at length. Also examined, in the final Chapter, are the ways in which these prospective technologies, deployed or not, are gendered. In what follows, however, I focus on applying Helen Longino's feminist contextual empiricist (FCE) approach to science to climate engineering in order to

© The Author(s), under exclusive license to Springer Nature Switzerland AG 2019   45
T. Sikka, *Climate Technology, Gender, and Justice*, SpringerBriefs in Sociology,
https://doi.org/10.1007/978-3-030-01147-5_3

solidify the argument that, on the grounds of her critical and comprehensive feminist approach to science, geoengineering fails to meet the feminist virtues she articulates. In making this argument, several unnoticed and obscured insights into the assumptions, values, structures and implications that underpin geoengineering are made manifest.

As stated in the introduction, I draw on stratospheric aerosol geoengineering specifically in these Chapters because of its conceptual boundedness, which is methodologically helpful with respect to the pure volume of research on this subject as well as the extent of scientific progress on the SRM technique; its relative familiarity amongst the public (via the media); and because concentrating on this case allows for deeper investigation, evaluation, and understanding. This method "provides opportunity…to gain a deep holistic view of the research problem, and… facilitate describing, understanding and explaining a research problem or situation (Baškarada 2014, 1; Tellis 1997; Dooley 2002) in manner that taking up geoengineering in its entirety does not.

The final Chapter, which discusses the contributions of FST and technofeminism, considers climate engineering technologies in a more general way – as an ideal type. This is in line with the objective of demonstrating specifically how both FST and technofeminism can contribute a set of central insights into the gendered implications of geoengineering that renders FCE more robust. As will be explained, there are a number of notable insights of significance that come from the analysis of geoengineering techniques outside of sulphate SRM which is why the range of analysands are expanded in these sections.

Feminist Contextual Empiricism

Essentially, the feminist empiricist approach, of which Longino is one adherent, combines feminist objectives with traditional empirical science. Its purpose is to transform how science is practiced by evaluating the relationship between values and evidence in order to "think through a particular field and try to understand just what its unstated and fundamental assumptions are and how they influence the course of inquiry" (Longino 1987, 62). Here, I provide an overview of Longino's approach beginning with a discussion of her stand on objectivity, pluralism, sociality, values, and feminist scientific virtues. This sets the stage for the analysis of solar climate engineering in relation to her feminist virtues which makes up the greater part of this book and which establishes the argument that, based on these virtues, this approach to climate mitigation is not desirable.

Longino

Helen Longino's feminist contextual empiricist (FCE) approach to science merges contextual empiricism with feminism and philosophy of science. On the level of empiricism itself, Longino retains a focus on sensory experience but also inserts an acknowledgement of the role theory selection can and does play in scientific processes. Theory selection itself is based on constitutive values as well as contextual values. The former refers to those values that are generated in line with reaching scientific truth or success (defined empirically), while the latter are derived form our socio-historical contexts and involve the desire, for example, for a more just world.

Constitutive values inform how we select theory, as well as what we consider significant in evidentiary terms, and are rooted in Kuhn's identified values of accuracy, consistency, broad scope, simplicity, and fruitfulness (Kuhn 1977, 74), while contextual values attempt to remove, for example, "structural and institutional barriers that prevent feminist from participating in and influencing the course of scientific enquiry" (Longino 2005, 215). While my focus is on contextual values and their role in explaining how the socio-cultural environment in which science is done influences outcomes, conclusions, and the direction of science, constitutive values are equally significant in determining what constitutes acceptable scientific practice (Longino 1990, 4).

Pluralism is another subject that runs through Longino's work and is significant in relation to geoengineering. Essentially, pluralism involves the active adoption of a stance of openness and constant revision. In practice, it emphasizes the central insight that scientific controversies are both "complicated and multifaceted" requiring the rejection of the monist assumption that a single approach to knowledge is either possible or necessary (Longino et al. 2006, xxv). In doing so, it creates an opening for new ways of thinking that can "be evaluated on the basis of what they help us understand, not on the basis of one single, essential way of understanding" (Longino et al. 2006, xxvi). Longino, in her article, "Pluralism and the Scientific Study of Behavior," gives the example of how the phenomenon of aggressive behavior has been explained by variety of disciplines, each with equally valuable perspectives. What a pluralist stance holds as significant is that that explanations based on genetics, environmental factors, personality traits, or brain development have all been ratified through intensive evaluation and debate by their knowledge communities. The question is not whether one approach is correct, but that, in a pluralist scientific environment, critical debate amongst advocates of each perspective should be fostered thereby enabling "the refinement of methodologies, the clarification of concepts…" which "…makes for more knowledge…" that is also "…better knowledge" (Longino 2006, 127). Longino also asserts that each approach yields partial accounts, which is to say that each theory produces "some knowledge of behavior by answering the questions distinctive of it with methods that are also distinctive" (Longino 2006, 127). This insight will prove essential in understanding how knowledge generated from geoengineering models, algorithms, computer simulations and physical laws can stand in equivalent relation to insights produced by other knowledge practices.

A further set of ideas relevant to geoengineering research are rooted in Longino's firm objection to value neutrality in scientific practice. The rationale for this position centers on the critical distinction she makes between the context of discovery and the context of justification. The former, which refers to the context in which the origin of theories and ideas emerge, is value laden and constituted not only by the "circumstances surrounding its initial formation – its origin in dreams, guesses and other aspects of the mental and emotional life of the individual scientist" (Longino 1990, 146), but also by the "social structure or socioeconomic interests of the context within which an individual scientist works" (Longino 1990, 64). The latter, the context of justification, expresses how and in what ways hypotheses themselves are tested and verified.

Yet, as Longino makes clear, in regards to the context of justification (and in light of the principle of underdetermination of theory by data), contextual values are not excluded. Her underdetermination thesis asserts that there is a gap between data, comprised of experiments and observations, and hypotheses, understood as statements about the data which "always exceeds that of the statements describing the observational data" (Longino 1990, 58). As such, FCE recognizes "not just the influence exerted by social and political factors in the justification of theories but a voluntarism whereby scientists may justifiably (without logical constraint) choose between two empirically adequate theories depending on which is most consonant with their values" (Gannett 2008, 10).

An example of this includes hypotheses about the effect of carbon emissions on future local air quality of or, in the case of SRM, how the size of sulphate particulates in the atmosphere might change over time. The logical gap between the data and theory is filled by background assumptions that decide what counts as evidence based on community norms.

However, this does not imply that, objectivity and truth is beyond reach for the reason that, in a community context, science is subject to criticism of evidence based upon shared standards. These standards include, but are not limited to, constitutive values like "empirical adequacy, truth, generation of specifiable interactions with the natural or experienced world, the expansion of existing knowledge frameworks [fruitfulness], consistency with accepted theories in other domains, comprehensiveness, reliability as a guide to action, relevance to or satisfaction of particular social needs" (Longino 1990, 155).

It also includes community responsiveness to criticism, which must be adopted, and a level of equality with respect to intellectual authority whereby minority and diverse voices are not silenced. This mode of transformative criticism constitutes conditions of objectivity defined as the "ideal by reference to which particular scientific communities can be evaluated." Longino asserts that the ability to ascertain the "practices and institutional arrangements that facilitate or undermine objectivity in any particular era or current field, and thus the degree to which the ideal of objectivity is realized, requires both historical and sociological investigation" (Longino 1990, 80).

It is in light of the fact that, knowledge production is social, and rigorously so, that the danger of idiosyncratic ideas are eliminated. Supporting this kind of objectivity is Longino's claim that equality of intellectual authority is constrained by both nature and logic (Longino 1993, 113). Relativism is thus circumvented as a result of the limitations that have been placed by community standards. What this does is to highlight the fact that "cognitive needs can vary and this variation generates [productive] cognitive diversity" (Longino 1993, 113). Having appropriate venues for debate; shared, socially-generated norms; a willingness to incorporate criticism; an openness to underrepresented groups and perspectives (including women); and, a degree of intellectual equality that ensures that underrepresented knowledge is respected, work together to "block influence of subjective biases in a way that is unachievable for an individual" (Eigi 2015, 451; Longino 1990, 2002). With respect to geoengineering, this leaves ample room to expand and enlarge an assessment and critique without undermining the principles of scientific objectivity or providing grounds from which to challenge climate science.

The feminist dimension of the FCE approach is both subtle and transformative. It is made manifest in a number of ways stemming from the central observation that interdependence, as a relational truth, is a particularly feminist insight (Longino 1993, 111). Longino calls this 'doing science as a feminist' which sets out a clear role for feminist politics in science, emphasizes science as a collaborative practice founded on "research programs that are consistent with the values and commitments we express in the rest of our lives" (Longino 1990, 191), and which "reveal both gender in the phenomena and gender bias in the accounting of them" (Longino 1997, 45).

Doing science as a feminist entails two types of action. First, on narrow level, it means that if we accept that the context of inquiry can shape how knowledge is produced, then it stands that "gendered values and standpoints – shapes direction and presuppositions of inquiry, and the content and evaluation of knowledge claims" (Wylie 1995, 346). This is what Longino's six feminist virtues are founded on – namely, making gender present alongside agreed upon evaluative scientific standards. It also means, on a philosophical level, that epistemic standards are open to constant revision. Wylie argues that, for Longino, doing epistemology as a feminist entails "a preparedness to modify the tools we use and the presuppositions we make both as epistemologists and as feminists" (Wylie 1995, 354).

Scientific Virtues

Longino's set of feminist values of inquiry are treated as a non-exhaustive range theoretical virtues that protect against ontological monism. As theoretical virtues, they are understood as "characteristic of theories, models, or hypotheses, that are taken as counting prima facie and ceteris paribus in favor of their acceptance"

(Longino and Lennon 1997, 21). They include the virtues of empirical adequacy, novelty, heterogeneity, mutuality or reciprocity of interaction, applicability to human needs and diffusion of power or universal empowerment and are posed in contrast with the commonly held scientific values of "simplicity, explanatory power and generality, fruitfulness or refutability" (Longino and Lennon 1997, 21). These values are not meant to replace Kuhnian ones, but, rather, to illustrate that there is more than one set of reasonable scientific standards. It also demonstrates how" feminist values fulfill a cognitive function in promoting specific (feminist) research goals" by illustrating how even the "Kuhnian values are possibly connected with social values" (Büter 2010, 223; Longino 1995, 1996). I briefly describe each of these feminist virtues below before conducting the analysis of SRM geoengineering in light of them.

For Longino, empirical adequacy is the only proper epistemic virtue of the six and refers specifically to the quality of fit between theory and observation (Longino 1990; McMullin 1983). Observational findings generated in the course of experimentation, as well as quantifiable operations, constitute its central criteria and are shared by feminists and non-feminists alike. Hugh Lacey, drawing on Longino, explains that empirical adequacy should, amongst other things, exhibit predictive power, be empirically testable, falsifiable, and contain rich informational content about significant empirical phenomena (Lacey 1997, 31–32). However, Longino asserts that this virtue alone cannot generate sufficient consensus. As such, she brings in other non-empirical criteria in order to generate a satisfactory level of epistemic acceptability.

The second standard, novelty, is part of the larger framework of understanding relied upon as well as serving as a counterpoint to conservatism in which the objective is to make visible "salient aspects of experience or reality hidden or marginalized by presently accepted theory" (Longino and Lennon 1997, 22). Heterogeneity, the third virtue, refers to a preference for multi-causality and a struggle against unified theories and explanations. Longino and Lennon note that feminist researchers have traditionally "resisted unicausal accounts of development in favour of accounts in which quite different factors play causal roles" (Longino and Lennon 1997, 22). Fourth, mutuality/reciprocity of interaction implies an inclination towards theories that are not only complex, but also represent phenomena as interconnected and thus engaged with diverse actors and entities. Longino gives the example of DNA which is commonly understood as a 'master molecule' acting as a 'master of causality' as one example of the type of scientific practice to be avoided (Longino 2008). Also of note is the virtue of attending to human needs. This particular value is foundational to the practice of a feminist science in which care, justice and privation should guide scientific objectives and decision-making. Also significant to this pragmatic virtue is Longino's assertion that these needs, both human and social, tend to be the ones most often ministered to by women (Longino 1996, 38). Finally, the last virtue, also pragmatic, is the decentralization of power, which asserts a preference for forms of knowledge and technological applications that challenge entrenched power relations and empower communities. As will be demonstrated, this is of particular importance to the subject of geoengineering. It also refers to institutions that produce knowledge, which should be diverse, open, and autonomous (Longino 1996).

Taken together, these seminal virtues work in tandem to guide scientific practice in a direction that is just, prosocial, and, most importantly, feminist. For Longino, this means that, these virtues are not *feminine*, but *feminist* in the sense that "If the context is gendered (in the sense of being structured by gendered power asymmetries), inquiry guided by these virtues is more likely to reveal it or less likely to preserve its invisibility than the traditional virtues" (Longino 1996, 50). In what follows, I use these virtues to reveal how gender operates within, amongst and around the practice of geoengineering (sulphate SRM specifically), and also establish that the scientific practices constitutive of climate engineering do not, on the whole, accord with these feminist principles. Taken together, this approach provides a much needed critique and assessment of geoengineering from a feminist perspective that has not been adequately attended to by existing research.

Empirical Adequacy: Virtue One

Empirical adequacy, as explained above, is the sole cognitive epistemic virtue articulated by Longino. On this level, SRM geoengineering can be analyzed from a variety of perspectives. First, is the basic science – i.e. is there a fit between data and the theories articulated and, if so, what is it based on? This requires an unpacking of the data itself, the theories produced, and the background assumptions that connect the two. Remember that there is an underlying thesis of incommensurability and underdetermination here such that Longino, like Kuhn, asserts that more than one theory can be derived from the same or similar data. As such, and with respect to SRM, this implies that the assumptions which lead to the conclusion that model-generated data supports the belief that sulphate climate engineering is a viable option, should also examine the assumptions that lead to the opposite conclusions.

It is also the case that, the majority of research into SRM geoengineering is conducted without the luxury of direct observables or historical data – meaning that it relies on a series of equations, models, and extrapolations to reach conclusions. This reliance on theoretical rather than observational information requires confidence in a set of assumptions that are embedded in a particular scientific framework. Unpacking this framework constitutes another significant are requiring further analysis. In order to execute this component, as well as study the assumptions built into the theoretical language used, it is useful to take on a seminal piece of research that actually employs the theoretical framework, models, simulations, and theories used to conclude that SRM is a feasible path to climate mitigation.

Before doing so, it is important to note that, in addition to reveal the operating theoretical assumptions and frameworks involved, Longino makes clear that the best way to develop a feminist science is not only to make these constituent elements transparent, but to also offer an empirically adequate yet pluralistic alternative as a feminist act. What makes this difficult in the context of geoengineering is that previous uses of this technique have relied on research conducted in the biological sciences, like that of gender role behavior – e.g. whether gender is hormonally determined or the outcome of more complex self-conscious, agential, and self-reflective analysis (Longino 2005), which lends itself to an explicitly gendered investigation.

This kind of analysis is considerably more complicated in relation to the natural or physical sciences which do not have an obviously gendered dimension. However, if we take Longino's assertion that displaying normative commitments and challenging fixed assumptions is a feminist act, thereby expanding human potentiality (Longino 2005), it is possible to conduct such an investigation on feminist grounds. Additionally, it is also the case that, as in the preceding example, conclusions reached on collected data can be justified in more than one way in accordance with the underdetermination thesis that lies at the heart of empirical adequacy. As such, it is important to point out that what constitutes an empirically adequate theory varies with context since the gap that exists between data and observation is always filled by social values and background assumptions which are not monocausal and can be explicitly feminist. Therefore, regardless of the lack of a specific thematization of gender, it is possible to take a feminist approach with respect to sulfate climate engineering – that is, to both uncover and disclose value-laden assumptions, offer alternative theories associated with the data, and pry open theoretical frameworks that offer static explanations. This is doing science as a feminist.

A revealing set of findings suitable for analysis can be found in the GeoMIP approach mentioned in the Chaps. 1 and 2. I focus specifically here on the published findings found in the article, "Climate model response from the Geoengineering Model Intercomparison Project (GeoMIP)" by Kravitz et al. published in 2013. Taking these published findings as cases for analysis, in line with the virtue of empirical adequacy, entails a study of the specific procedures used in the course of experimentation as well as the conclusions and theories extrapolated from them. What is most probative here are the assumptions and values that fill the gap between data and conclusions. After establishing this, an analysis of what these values reveal socially and politically, as well as how they fit with the dominant epistemological framework, is conducted followed by an exploration of whether these are the only values that can explain this connection. In the course of this analysis, the subjects of modeling, data-theory relationship, boundary objects, visualization and language are considered.

The decisive finding from the GeoMIP project is that, based on 12 climate models run under a variety of scenarios, SRM climate engineering using sulfates is a feasible option in mitigating climate change given careful planning and further research. Kravitz et al. assert that the results of these model simulations, taken as a whole, is that in the case of a quadrupling of CO_2 from preindustrial levels, stratospheric sulphate geoengineering could be expected to bring CO_2 levels into balance – which is to say that under each of the scenarios set out in the experiment, sulfate climate engineering "largely offsets global mean surface temperature increases due to quadrupled CO_2 concentrations" (Kravitz et al. 2013, 2320). This finding is based on data generated from highly complex simulations with multiple data-related decisions being made regarding, for example, the control which, in this case, is a preindustrial level of CO_2 concentration of 280 ppm; time scales, which vary; differences in abruptness of radiative forcing; and the inclusion of additional parameters that test for changes in sea ice, precipitation, and the effects on clouds.

Again, the focus here is not to question the veracity of the conclusions reached as a result of the complexity of the science (it in some ways reflects the kind of feminist pluralist science supported by FCE). Rather, the examination of the virtue of empirical adequacy with respect to sulfate geoengineering is aimed at engaging with the values and assumptions used to make the data-conclusion link possible in light of these complexities. The choice of preindustrial CO_2 levels as the control is just one of several value laden choices being made that links collected data to reached conclusions. I begin with an examination of this assumption below.

Boundary Object and Baselines

The choice of preindustrial CO_2 as the baseline in the GeoMIP project is significant because it assumes, appropriately, not only that human intervention is to blame for climate change, but that this disruption can be traced back to a definable set of socio-historical changes that began with industrialization. Industrialization is marked, amongst other things, by the rise of urbanization, mass factory production, the use of coal and steam for energy, as well as changes in communication and transportation technologies that, together, have markedly increased levels of CO_2 emissions.

In scientific experimentation, choosing a baseline is a normative act and opening it up to the question of why this particular baseline (280 ppm), and to what end, are of utmost importance in undertaking a feminist interrogation of accepted science. One of the ways in which baseline assumptions can be investigated is by thinking about how they function as boundary objects. The central function of a boundary object is to act as a mechanism for knowledge translation and mediation between experts and the public – which is essential in "developing and maintaining coherence across intersecting social worlds" (Star and Griesemer 1989, 393). Understood this way, the production of such an object can be seen as a outcome of careful intellectual work that acts as a substantive factor in extrapolating collected experimental data to the reached conclusion that sulphate climate engineering is a reasonable option.

Moreover, this particular boundary object discursively and methodologically assumes that CO_2 is the primary driver of climate disruption which ignores, for example, the effects of methane and NO_2 emissions (Howarth et al. 2012; Seinfeld and Spyros 2016). It also overlooks the fact that early activities such as deforestation due to agricultural activities (e.g. hunting and gathering) and the transition to livestock rearing were the earliest drivers of human induced climate change – albeit on a smaller scale (Kaplan et al. 2009; Olofsson and Hickler 2008). Also of note is recent research suggesting that mass extinction of prehistoric animals, like

Australia's giant kangaroos and marsupial rhinos and leopards, were a result over-hunting by humans centuries prior to industrialization (University of Exeter 2008) (Faurby and Svenning 2015; Braje and Erlandson 2013). The point here is that the pattern of harmful intervention by humans into the environment for survival, and then subsequently for economic growth, is not new. Yet, the 280 ppm baseline implies that it is by suggesting that that were we to get back to that preindustrial level, via geoengineering, the fundamental problem would be solved.

Chris Turney, lead author of the Exeter Report on climate, echoes this sentiment, asserting that, "It is sad to know that our ancestors played such a major role in the extinction of these species – and sadder still when we consider that this trend con-tinues today" (University of Exeter 2008). This lacunae in the conclusions reached by the GeoMIP report is essential in identifying significant underlying assumptions with respect to how empirical adequacy is theory laden and, as such, how largely positive conclusions have been reached with respect to sulphate climate engineer-ing's efficacy. Longino is clear that, in all cases, empirical adequacy requires this kind of interpretation such that it can be "meaningfully employed in a context of theory choice…" and whose "interpretations are likely to import socio-political or practical dimensions" (Longino 1995, 395).

Furthermore, the 280 ppm baseline does more than simply fulfilling these condi-tions as a boundary object. It also represents an assumption of cultural significance in which the pre-industrial and pre-capitalist period is seen as a comparatively pris-tine moment in time in which humanity's disruption of the natural world was not life threatening. A modern version of this, relevant to the case of geoengineering, is the critique of developmentalism which questions the ethos of industrialization as it relates to the destruction of local cultures, enhancement of wealth extraction, distor-tion of traditional practices of accumulation, and promotion of extractive economic policies that disrupts cultures, communities, nature and politics (Bergesen and Boswell 2000, 94) (Rueschemeyer et al. 1992; Seidman 1983). It is clear, from the evidence given above, that a clear distinction between a pristine preindustrial/pre-capitalist society and a destructive capitalist one cannot be maintained – yet, the presence of this assumption in culture, social theory and philosophy is significant in shaping how scholars and scientists set baselines with respect to geoengineering research as well.

Finally, it is important to recognize that the use of boundary objects comes with its own set of limitations and difficulties. This includes the possibility that alterna-tive viewpoints are constrained, thereby leading to selective use of data. In the case of solar geoengineering, an argument can be made that reliance on 'global average temperature' ignores local differences which, in turn, shapes our priorities around policy formation. It is also the case that CO_2 baseline fails to recognize the presence of the lags in the climate system we are likely to face as a result of the irreversible impacts climate change has had on our ecosystem (e.g. species extinction). Boundary objects can also be criticized as limiting alternatives by solidifying one particular point of view and negating "active negotiation of shared understanding" (Lee 2007, 313) at the outset. Understanding that these limitations exist in the center of the data-theory nexus brings us closer to a more comprehensive understanding of the auxiliary assumptions that make this cognitive connection possible and, consequently, a belief in the empirical adequacy of the conclusion that sulphate geoengineering can result in reduced global warming.

Now, it is important to note that the Kravitz et al. GeoMIP article itself is replete with research related caveats including that their findings are limited by idealizations and that they have been unable to account for extreme events or ocean acidification in the model. The authors go so far as to assert that,

> The results we present are specific to the highly idealized Experiment G1 and should neither be mistaken as an evaluation of geoengineering proposals or issues surrounding their implementation, nor as representative of how geoengineering might be implemented in practice (Kravitz et al. 2013, 8321).

Yet despite these qualifications and stipulations, Kravitz et al. are comfortable enough in their findings to maintain that an overall lowering of global climate is likely. It is important to note that the prominence given to this central finding is significant as it is the overriding objective of the entire enterprise.

Modeling, Idealization, and Parameterization

In additional to boundary objects, idealizations, parameterizations, and variable choice constitute further significant factors that play an important role in guiding knowledge/theory formation, decision-making, and thus in establishing empirical adequacy. Longino argues that idealizations are established by value judgments in ways that shape conclusions and theory choice. Her most persuasive case study has to do with how the leading studies of aggression rely on idealizations based on genetic rather than epigenetic or intrauterine causality (Longino 2013).

Understood in this way, because of variable choice, the interpretation of underlying causal processes based on idealizations and parameterizations can be thought of as underdetermined and thus open to alternative interpretations. One of the key parameterizations built into the GeoMIP model is the elevation of the global over the local. On the first page of the article it is made clear that its priorities are gaging "globally uniform reduction in insolation," "global mean surface temperature" and "global average net primary productivity" (Kravitz et al. 2013, 8320). This focus is clearly the case notwithstanding Kravitz et al.'s assertion that although global mean temperatures would be offset by geoengineering, it is also likely that tropics would be cooler and receive less precipitation, polar regions would be warmer, and, "that uniform solar geoengineering…cannot simultaneously return regional and global temperature and hydrologic cycle intensity to preindustrial levels" (Kravitz et al. 2013, 8330). Despite these qualifications, thematizing global over regional climate in the piece underscores the conclusion that "globally averaged surface air temperature may be kept at preindustrial levels" (Kravitz et al. 2013, 8330), through sulphate engineering which makes it possible to avert the worst consequences of global warming.

Not only does the prioritization of the global over the local constitute a significant act of parameterization, but it also establishes a set of prescriptive value judgements that lend further support to the empirical adequacy of the positive conclusion reached on sulphate geoengineering. Briefly, this binary is rooted in what can be thought of as a 'politics of scale' (Smith 2010) wherein a Western and neoliberal ontological preference for globalization is entrenched and expressed in scientific assumptions as well as political ones. I return to this binary and explore its relevance in much more detail in the section on heterogeneity.

Some further discussion of model choice with respect to empirical adequacy is also warranted. Within feminist science, multiple models that ask different questions and highlighting diverse factors are encouraged. And on this point, variation, of a particular sort, is not lacking in SRM geoengineering research. Over the past several years a number of geoengineering intercomparison projects have been initiated with the objective of attaining more robust, consistent, and representative results. Many of these use ensembles of simulations that manipulate variables, employ perturbations, and utilize different initial states – yet, they are largely consistent in their conclusion that incoming radiation, and therefore temperature, would be reduced were solar climate engineering deployed (Rasch et al. 2008; Robock et al. 2008; Berdahl et al. 2014). While the significance of epistemological diversity and modeling choice is an area of research noteworthy in and of itself, the objective of this book is not to engage deeply with the minutiae of scientific modeling, but to establish how, and on what assumptions, climate modeling works to establish empirical adequacy.

Briefly, it is important to point out that, climate models do not simulate future climate but produce relative probabilities of potential future climate states. Faith or trust in these models, as well as their empirical adequacy, is generated through the process of discursive consensus formation articulated by Longino – which is discussed further below. Typically, it is General Circulation Models (GCMs) that are employed in climate science. These models, because they do not use direct empirical data, establish plausible scenarios rather than determined end states. As such, they "are treated as a kind of intermediary – between investigations and the world in their role as indirect representations, and between explanations and the world insofar as they are used to generate understanding" (Potochnik 2012, 385; Woodward 2003).

The predominant model currently preferred for research into geoengineering is a specific type of GCM, atmospheric and oceanic general circulation models (AOGCM), which divide the ocean and atmosphere into a three dimensional grid that begins with baseline state and incrementally moves forward in time. Multiple variables from the carbon cycle to vegetation, temperature, clouds, precipitation, sea ice etc. are built into the model with an emphasis on oceanic and atmospheric feedbacks. With respect to sulphate climate engineering, these processes are particularly significant as they determine phenomena like precipitation levels. It is also the case that an increase and/or decrease of solar radiation indicate *the* fundamental variable of this research. On this point, AOGCM models have shown themselves to be particularly adept (Robock 2014a, b; Llanillo et al. 2010). Moreover, they are consistent in concluding that the injection of SO_2 into the stratosphere can counteract GHG-induced warming (Jones et al. 2010; Caldeira and Wood 2008; Lunt et al. 2008). It should be noted, however, that the data which generates this conclusion relies on physical parameterizations made by scientists that aim to account for details, such as including reflectivity, clouds, and turbulence, that cannot all be represented in the model due to problems of scale and complexity. It is notable that the trend in modeling, for which Kravitz et al.'s GeoMIP experiment is just one example, has been to include as many factors as possible – e.g. coupled models (of which AOGCMs are one example) – which may end up introducing more chances of error and uncertainty.

This is because a highly complex model is positively correlated with low resolution and an increased chance of error (Edwards 1999, 442). This inclination towards high density reflects a cultural preference in scientific practice for complexity, in the form of additional parameterizations, which is often not necessary. This is particularly the case since if the core purpose is to model solar forcing, as in the case of solar geoengineering, less complicated models like energy balance models that captures essential large scale features are perfectly acceptable (Shackley 2001; McInnes 2010). More generally, if the empirical adequacy of the supposition that solar geoengineering is viable relies on parameterizations, it is essential to unearth the assumptions on which these choices are made – while keeping in mind, of course, the modest, and not necessarily predictive, contribution these models make with respect to scientific knowledge.

This is an area in which, gender is explicitly manifest in the preference for big data which associates more data with higher intelligence, accuracy, objectivity and truth. This inclination is rooted in a history in which 'Big Science' is simultaneously correlated with control, militarism, aggression, and a generalized masculine culture (Wajcman 1991, 143). Not only is the focus on objectivity that stems from this privileging of big data often ontologically unsound, but it can also result in a kind of scientific practice where,

> …the choice of topics often implicitly supports sexist values; female subjects are excluded or marginalised; relations between researcher and researched are intrinsically exploitative; the resulting data are superficial and overgeneralised; and quantitative research is generally not used to overcome social problems (Oakley 1998, 709).

In privileging more data, coupled models, multiple parameterizations and perturbations, climate models, such as the one used in the GeoMIP article, can be critiqued as following in the gendered footsteps of its predecessors. This is not, however, an argument against complexity – which is in fact one of Longino's central virtues – but a questioning of the 'bigger is better' ethos that dominates traditional science.

While a closer examination of the precise mathematical principles, parameterizations, and induced perturbations used to reach the conclusion that SO_2 climate engineering is likely to be effective is beyond the scope of this book, it does constitute an area for further research. There are, however, a number of further model related assumptions worth discussing in the context of feminist science and empirical adequacy. First, the images, graphs, and charts produced from these model-based experiments are significant cultural objects in that they contribute to the belief in empirical adequacy which, being rooted in the data-theory relationship, is necessary to establish trust in knowledge. They also function as visualizations that "muddle the lines separating number and figuration and are at the heart of climate and policy making" (Houser 2016, 16). Consequently they act as further examples of background assumptions and value judgments built into the space between data and theory. This subject is discussed in more detail in the Chapter that examines the virtue of heterogeneity with respect to alternative modeling and representational practices.

Visualizations

Another way in which the empirical adequacy of the claim that, based on data, sulphate climate engineering can potentially mitigate global climatic warming can be assessed is by investigating the role played by images in reinforcing this assertion. Heather Houser (2016), in her article "Climate Visualizations as Cultural Objects," argues that climactic visualizations are both artistic artifacts and cultural objects that shape our understanding of climate change and does so in relation to geoengineering as well. She is clear that because these rhetorical objects are meaning-laden, and thus not value neutral, they constitute a significant site of investigation.

Schneider echoes this assertion by explaining how these images not only impart information, but also represent a "paradigmatic field in which images take on the role as political agents" (Schneider 2012, 188). As such, everything from the temporal scales and color choice to the cartography produces knowledge (Houser, 142). Figure 3.1, from the Kravitz et al.'s GeoMIP article, is a prime example of how images can function as agents and visual containers used to establish the empirical adequacy of their lowered mean surface temperature assertion. What follows are selected insights and analysis of this particular visualization which functions as a paradigmatic case, especially with respect to the role it plays and the values it reflects, in establishing empirical adequacy.

Figure 2.2, it is asserted, represents "all model ensemble average surface temperature differences" for abrupt and controlled (geoengineered) climate change. The three 2D images of the world on the left represents the state of affairs in the event of a quadrupling of CO_2 emissions on an annual average basis (top left) as well as the averages for the months of December, January, and February followed by June, July, and August directly below. On the right are the same 2-D images representing the climatic state of affairs if sulfate solar geoengineering were to be implemented. What is immediately striking about these images is the use of color. Most notably, the figures on the left are colored a bright red, indicating a high degree of warming.

KRAVITZ ET AL.: GEOMIP MODEL RESPONSE

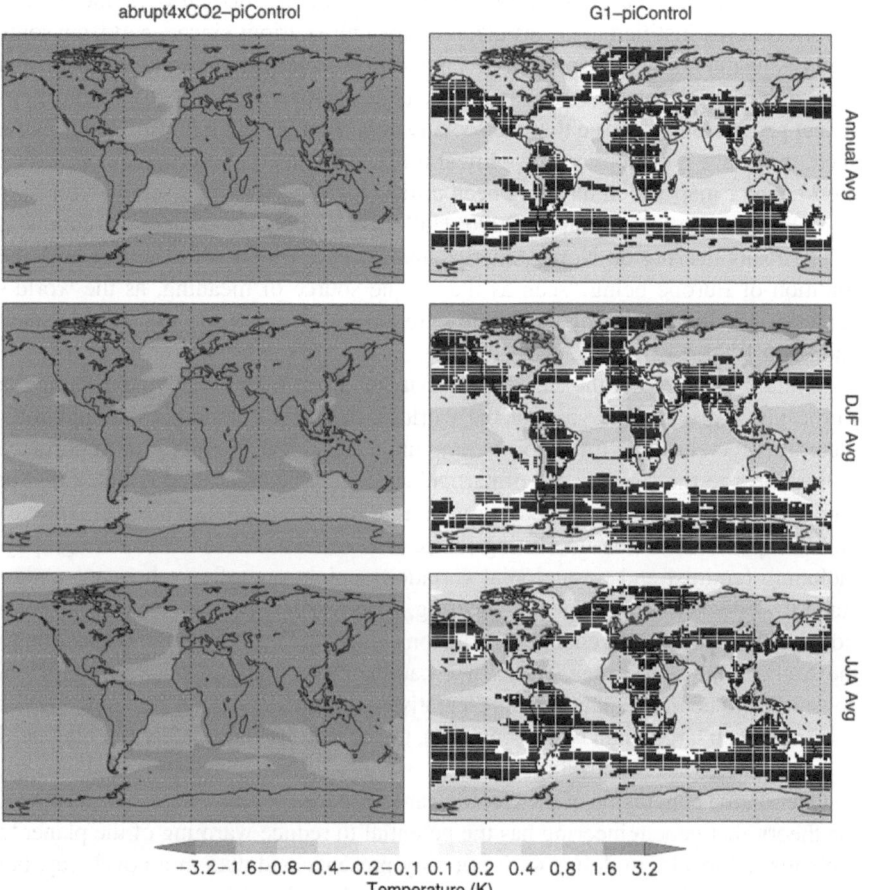

-3.2 -1.6 -0.8 -0.4 -0.2 -0.1 0.1 0.2 0.4 0.8 1.6 3.2
Temperature (K)

Fig. 3.1 Ensemble representation of differences between air surface temperature if geoengineering is implemented. (Kravitz et al. 2013, 8323)

While this has become a customary way to represent heat, it is also "culturally symbolic, the red, rising end of a temperature curve or red coloring of a globe can be perceived as a modern way of picturing disaster" (Schneider 2012), and emergency in a manner that is epistemologically forceful. The choice of the much less striking palates of light oranges to white and then blues is equally visually compelling in their association with tranquility and calm (Stone and English 1998; Kress and Van Leeuwen 1996). The impact this has on a reader can be compelling by generating a sense of receptiveness through the dichotomy established between danger and safety.

Also of note is the fact that these images are 2D and rectangular, which is unusual in that most visualizations tend to either be ellipsoidal, which shows the area of the continents and thus temperature rise distribution more accurately, or as 3D spherical impressions which are considerably more compelling. This cartographic choice, never explicitly justified in the article, is all the more curious since, apart from some slight adjustments (e.g. a smaller Greenland than usual), it employs an antiquated map of the world. This is significant since, as Klinghoffer argues, "geographic maps reflect perceptions of space that are socially conditioned" (Klinghoffer 2006, 7). As such, they have a symbolic and political impact. The map chosen here is based on the Mercator map which has long been criticized by postcolonial scholars for having a Eurocentric bias (Johnson 2006; Hall 2014). This map centers the effects of solar geoengineering on North America and Europe, which is in keeping with the tradition of Europe being "seen as the unique source of meaning, as the world's center of gravity, as ontological "reality" to the rest of the world's shadow" (Shoat and Stam 1994, 1–2).

While it is likely not the case that Kravitz et al. are intentionally suggesting the Eurocentric cartographic view of the world is the most comprehensive and complete, it is the case that this choice communicates a clear hierarchy of importance with respect to the significance of particular geographic areas, and populations for whom the effects of geoengineering would be most felt – namely North America and Europe. It would be interesting to see how more novel forms of cartography, including feminist and postcolonial variations, might reconfigure how we understand the effects of sulfate geoengineering, with respect to their visual representation, and how this might change conclusions reached. When posed in opposition to the AuthaGraph world map, (http://www.authagraph.com/top/?lang=en), which is considered to be more spatially representative – with Africa taking on a much larger size (i.e. much larger than North America), the contrast with the Mercator map used in the GeoMIP piece is striking.

Overall, the conclusion of the GeoMIP article, based on this visual data, advances the theory that geoengineering has the potential to reduce warming of the planet to preindustrial levels. In doing so, it not only presents said data in a novel way, but also produces a further set of assumptions and values that fill the data-theory nexus required to establish empirical adequacy. Yet, it is also important to underline that this visual pluralism, and its intersection with representational choice, imagination, and political values, should not lead to a denial of climatic imageries as imaginary. Rather, confidence in models and visualizations has been achieved through accumulative discourse between scientists in which these images play a significant part. The value of these visuals is not only in their status as a source of data used to reach conclusions about the efficacy of sulphate geoengineering in their own right, but, suggestively, as artistic artifacts that are also value-laden representations with their own set of assumptions used to fill the gap between data and theory. As such, they have significant and robust epistemic and social authority.

Natural Analogues, Language, and Discourse

Finally, sulphate climate engineering is unique in comparison to other SRM and CDR techniques in that there are natural analogues available to further inform conclusions about likely outcomes. That is to say that these analogues, primarily volcanic eruptions, can and have been used to further reinforce confidence in the conclusions reached by models about the likelihood that sulphate climate engineering could reduce average global climate (Robock et al. 2013; Trenberth and Dai 2007). This poses a significant advantage to other climate engineering techniques for which novel climates and other ecological surprises may be more difficult to predict (William et al. 2007). In the case of temporal analogues, based on past evidence, as in the data collected documenting the cooling that followed Mount Pinatubo, they can make scenario construction more robust, provide much needed baselines, and, because changes were observed, yield results that are likelier to be both "internally consistent and physically plausible" (Mearns et al. 2001, 748; Pittock 1993). Yet analogues are also imperfect and, as Robock et al. argue, there are significant problems with relying on the volcanic analogue including the differences in the duration of effects, problems replicating particle size, and the uncertainties with respect to the effects of ash (Robock et al. 2013, Soden et al. 2002). As these observations demonstrate, even when analogues can be relied upon, the gap between data and theory is still replete with background assumptions and values that, consistent with Longino's feminist approach, warrants some scrutiny. I discuss this in more detail in the Chapter on heterogeneity.

An examination of discourse is also important in this context since the meaning of key terms and concepts that surround SRM geoengineering are often shaped by accepted theory. As such, it must be scrutinized critically so as to ensure that data is not being made to fit existent theoretical frameworks, which may contain gender distortions, and to better understand what assumed assumptions demand. A useful example that Longino uses is with respect to the term 'mass' which she argues means entirely different things in classical Newtonian mechanics, where mass is the property of a body, as opposed to relativist mechanics in which mass depends on the velocity of the observer. This, in turn, means that observations themselves are shaped "at least in part, by theory and described in language whose meaning was dependent on the whole of theory" (Longino 1990, 26–27).

In the context of SRM geoengineering, the choice of language is a significant indicator of the values being used to bridge data with theory. While a more fulsome discussion of discourse as a political enterprise is undertaken in the section on technofeminism, it bears some discussion in the context of empirical adequacy as well. While the examples of linguistic re-framing do not feature in the Kravitz et al. article, discourse deserves discussion as a larger phenomenon that shapes empirical adequacy with respect to how conclusions about SRM are reached and concretized. A notable linguistic modification illustrating this phenomenon is the recent rebranding of SRM as 'Sunlight Reflection Methods,' which has a much less negative connotation than solar radiation management. The former phraseology is much softer and less invasive (management versus modification or engineering), while simultaneously communicating an active cooperation with, rather than manipulation of, existing natural processes like sunlight. Similarly, terms like albedo modification, thereby eliminating any mention of sulphates, are also popular (Lenton and Vaughan 2009).

A further discursive change that has occurred consists of the retreat from using the term 'engineering' or 'modification' to describe these climate activities and, in its place, using 'climate remediation' (Sarewitz 2011). The cognitive baggage associated with the term engineering and modification tends to be interpreted as interventionist, risky, manipulative, and interfering with natural patterns. Climate remediation, on the other hand, communicates the objective of reversing climate change with tight regulation, as per the history of environmental remediation in the case of pollution.

Truth, Trust, and Consensus Formation

Regarding Kravitz et al.'s GeoMIP article, and specifically with respect to its central assertion vis-à-vis the possible success of sulphate geoengineering induced climate mitigation, it is important to highlight and unpack the process by which scientific consensus has been reached. For Longino, the prerequisite conditions necessary to attain scientific standing, and thus empirical adequacy, includes: spaces to engage in critical debate, openness to criticism, incorporation of said criticism, public standards, and equality with respect to authority. Objectivity and trust in conclusions results from engagement in this scientific practice, which is a product of the "critical interaction of different groups and individuals with different social and cultural assumptions and different stakes." According to this perspective, it is the "irreducibility of the social components of the scientific situation...that are, in fact, an essential part of the picture of scientific practice" (Lloyd 1996, 100). Kravitz et al.'s GeoMIP text has clearly gone through the process of peer review and, as a project comprised by the participation of multiple actors, is collaborative. However a few clarifications are necessary before a sense of the GeoMIP's dynamics and fidelity to a process of consensus formation can be ascertained.

It is important to point out that because Longino maintains that the truth value of scientific conclusions are "not opposed to social values...indeed it is a social value...whose regulatory function is directed/mediated by other social values operative in the research context" (Longino 2004, 135), she is able to hold on to an understanding of science as plural and value-laden as well as objective as long as conditions are in place for ideas to be modified, eliminated, and improved through a process of discursive interaction. In the context of SRM geoengineering, there appears to be a great deal of debate in designated venues. A significant component of establishing empirical adequacy has to do with the precise processes by which that adequacy is established. In the case of sulphate climate engineering, this component is formalistically fulfilled, but lacking in significant areas.

Within the solar geoengineering scientific community, there are recognized venues meant to facilitate discussion (the role of the wider community, both academic and non, are discussed further on). These include well-respected academic journals like *Climate Change, Atmospheres,* and *Nature, Environmental Research Letters,* and *The Royal Society* that have published numerous pieces on the subject. There is also an active conference circuit in which scholars, both advocates and detractors, come together to present ideas, perspectives, and new research. The 2010 Asilomar Conference On Climate Intervention Technologies is a notable example, as is the 2012 Geoengineering our Climate Conference in Ottawa, and the 2017 Gordon Research Conference on Climate Engineering in Maine. Geoengineering specific papers have also increased in generalized climate change conferences as have the publication of entire books dedicated to the subject (Burns and Strauss 2013; Blackstock and Miller 2016; Morton 2015). The GeoMIP project is itself a collaboration between a number of groups from Paris and Beijing to Potsdam with the findings and models made public.

With formal venues for discussion established, for empirical adequacy to be reached one also requires a openness to oppositional points of view resulting in an uptake of critical discourse. Longino asserts that, "The community must not merely tolerate dissent, but its beliefs and theories must change over time in response to the critical discourse taking place" (Longino 2001, 129–139). While criticism appears widespread, with detractors and opponents of geoengineering participating in conferences, journals, and books, the actual uptake of criticism is more difficult to ascertain. That being said, the fact that the latest research on sulphate geoengineering has begun to address issues of pluralism, risk, ethics and policy more directly suggests that things are moving towards that direction.

Public standards, as it relates to establishing empirical adequacy, is the third of Longino's criteria of public discourse which does appear to exist quite strongly within the geoengineering community. Amongst those studying sulphate geoengineering, modeling in particular is an area in which, while disagreements exists about the superiority of certain models over others (e.g. GCMs versus AOGCMs versus Multi-Model Ensembles in their various iterations), their usefulness and scientific value are not questioned (Jacob et al. 2007; Flato et al. 2013). Proponents also share a common vocabulary (see discussion of framing and discourse), and generally agree on the data as well as conclusions reached. This general confidence in modeling is based on it being derived from physical laws, having been successful in simulating aspects of the climate and physical processes, and being subject to comprehensive public tests (Randall et al. 2007; Stainforth et al. 2007; Knutti and Sedláček 2014).

In constructing (AO)GCMs, Coupled Global Climate Models as well as Earth System Models) used to simulate solar geoengineering, this confidence is dependent on assurances that the scientists involved in making judgments, with respect to parameterizations, perturbations, and basic model construction, are drawing on values that have been agreed upon – which it appears they have (Pawson, et al. 2008;

Lunt et al. 2008). However, it is important to note that public standards must also extend to the larger community of supporters, e.g. think tanks (AEI and the Heartland Institute), mainstream media (Hansson and Anshelm 2016), and notable private citizens (Bill Gates, Richard Branson), for whom common values and objectives are shared and self-reinforcing. This condition is also fulfilled judging from the myriad of individuals and entities who cite and deploy the results of scientific conclusions in their own work.

Finally, we have tempered equality of intellectual authority in which all participants are seen as capable contributors irrespective of class, gender, race, and/or socio-economic status. This does not mean anything goes – Longino is clear that this "equality must be qualified or tempered" (Longino 2002, 131) – but that compulsion, force, and marginalization are eliminated from scientific praxis. Findings that are judged to be empirically adequate depend on this being the case. This is where solar geoengineering runs into problems. With respect to representation in a general sense, there are a number of scholars from think tanks and research institutes, like The Kiehl Earth Institute and The Tyndall Center for Climate Change Research as well policy makers skeptical of geoengineering, for whom critical discussion is vibrant. An interesting set of critical debates are also occurring online amongst research groups and activist collectives like the ETC Group and Geoengineeringwatch, as well google groups, like geoengineering@googlegroups. com, which contain a rich store of analysis and research on all forms of geoengineering.

Yet, it is also the case that these conversations (amongst scientific supporters as well as critics of geoengineering) are taking place in discrete silos without significant cross discussion and debate. Conversations between scientists working on physical science, modeling, simulations, with social scientists studying issues around policy, ethics, justice and governance (Shepherd 2009; Preston 2013; Gordijn and Henk 2012) are few and far between. This lacuna is troubling in that it severely limits the kinds of discussions that can take place and forms of novel knowledge that can be produced. Finally, on the level of representation, there is also an unsurprising lack of of women and minorities in these discussions with respect to scientific practice. Engineering, computer science, and applied climate science in general have historically suffered from an absence of female representation for a variety of reasons that includes, but is not limited to, socialization, cultural norms, static educational practices, hostile work environments, and inflexible work schedules.

This is mirrored in work on geoengineering which, taken together, and when coupled with a Western, white male bias, means that the participation of women, racialized minorities, and peoples from the Global South is significantly lacking (Blickenstaff 2005; Hill et al. 2010; Hewlitt et al. 2008). This gap violates the equality of intellectual authority ethos on account of this two-track mode of knowledge production and the exclusion of significant social groups in geoengineering science. Empirical adequacy requires that the values that validate the connection between theory and data be grounded in an inclusive process of discursive consensus formation. Much more work needs to achieve this – particularly since, without it, what is lost are the very processes that "ensure epistemic effectiveness" through "a diversity of perspectives and dissenting opinions" that lead to a "pluralist view of knowledge" and which "offer[s] a more complete and textured representation of a problem or phenomenon" (Choo 2016, 112).

Overall, doing science as a feminist is precisely what Longino asserts examining empirical adequacy along these lines entails. Again, this does not involve a movement towards the pursuit of scientific practice that reflects 'feminine' norms and values, but, rather, ones that are "consistent with the values and commitments we [as women] express in the rest of our lives" (Tuana 1989, viii). Her democratic model of science incorporates regulative norms and an underlying fallibilism, localism and antifoundationalism that leave the door open for feminist issues and concerns. On the level of empirical adequacy, it remains possible to take an explicitly feminist stance with respect to, for instance, the status afforded to 'big science' and environmental science. Overall, an analysis of the virtue of empirical adequacy, as articulated by FCE, can only be conducted by unpacking the values and assumptions that connect data with theory in manner that complies with scientific standards set out by the communities working on sulphate geoengineering. A complete and self-critical assessment of scientific norms, values and assumptions that fill in and work to connect collected data with conclusions reached is precisely what I have done in this Chapter. The act of unpacking these assumptions, whether they explicitly thematize gender or not, constitutes doing science as a feminist.

Moreover, I have also illuminated and unpacked some of the most significant values with respect to modeling, visualizations, boundary objects, and the discourse relied upon to connect data with theory consistent with Longino's approach. The takeaway from this analysis is that there are indeed values embedded in scientific practice in ways that have often been left unexamined by applied science and solar geoengineering is no exception. Performing this analysis does not undermine the veracity of the science but lays bare the ways in which scientific practice can be improved and insidious values, including racist and misogynistic ones, eliminated. While this Chapter is lighter on critique, as far as empirical adequacy goes and particularly with respect to its aim to reconcile theory with observed data, it aims to be illuminative. In the following Chapters, the remaining feminist theoretical virtues of ontological heterogeneity, complexity/mutuality of interactions, applicability to human needs, and decentralization/universalization of power generated by engaged communities are drawn upon to further unpack the assumptions, virtues, and values that underpin sulphate climate engineering. These virtues are used to engage more critically with solar geoengineering with respect to substantiating my argument that, as it stands, this approach to climate mitigation should be rejected on feminist grounds.

References

Baškarada, S. (2014). Qualitative case study guidelines. *The Qualitative Report, 19*(40), 1–18.

Bergesen, A., & Boswell, T. (2000). *A world-systems reader: new perspectives on gender, urbanism, cultures, indigenous peoples, and ecology.* Lanham: Rowman & Littlefield.

Berdahl, M., et al. (2014). Arctic cryosphere response in the geoengineering model intercomparison project G3 and G4 scenarios. *Journal of Geophysical Research-Atmospheres, 119*(3), 1308–1321.

Blackstock, J., & Miller, C. (2016). *Geoengineering our climate?: Ethics, politics, governance.* London: Earthscan.

Blickenstaff, C. J. (2005). Women and science careers: Leaky pipeline or gender filter? *Gender and education, 17*(4), 369–386.

Braje, T. J., & Erlandson, J. M. (2013). Human acceleration of animal and plant extinctions: A Late Pleistocene, Holocene, and Anthropocene continuum. *Anthropocene, 4*, 14–23.

Buck, H. J., Gammon, A. R., & Preston, C. J. (2014). Gender and geoengineering. *Hypatia, 29*(3), 651–669.

Burns, W. C. G., & Strauss, A. L. (2013). *Climate change geoengineering: Philosophical perspectives, legal issues and governance frameworks.* Cambridge: Cambridge University Press.

Büter, A. (2010). Social objectivity and the problem of local epistemologies. *Analyse & Kritik, 32*(S), 213–230.

Caldeira, K., & Wood, L. (2008). Global and arctic climate engineering: Numerical model studies. *Philosophical Transactions. Series A, Mathematical, Physical, and Engineering Sciences, 366*(1882), 4039–4056.

Choo, C. W. (2016). *The inquiring organization: How organizations acquire knowledge and seek information.* Oxford: Oxford University Press.

Dooley, L. M. (2002). Case study research and theory building. *Advances in Developing Human Resources, 4*(3), 335–354.

Edwards, P. N. (1999). Global climate science, uncertainty and politics: Data-laden models, model-filtered data. *Science as Culture, 8*(4), 437–472.

Eigi, J. (2015). On the social nature of objectivity: Helen Longino and Justin Biddle. *Theoria, 30*(3), 449–463.

Faurby, S., & Svenning, J. C. (2015). Historic and prehistoric human-driven extinctions have reshaped global mammal diversity patterns. *Diversity and Distributions, 21*(10), 1155–1166.

Flato, G., et al. (2013). Evaluation of Climate Models. In: Climate Change 2013: The Physical Science Basis. Contribution of Working Group I to the Fifth Assessment Report of the Intergovernmental Panel on Climate Change. *Climate Change, 2013*(5), 741–866.

Gannett, L. (2008). *Why discovery matters: Values all the way down.* Philosophy of Science Association, Pittsburgh. http://citeseerx.ist.psu.edu/viewdoc/download?doi=10.1.1.522.4471&rep=rep1&type=pdf. Accessed 1 Nov 2017.

Gordijn, B., & Henk, T. H. (2012). Ethics of mitigation, adaptation and geoengineering. *Medicine, Health Care and Philosophy, 15*(1), 1–2.

Hall, P. A. (2014). *Counter-mapping and globalism. Design in the borderlands.* Abingdon: Routledge.

Hansson, A., & Anshelm, J. (2016). Has the grand idea of geoengineering as Plan B run out of steam? *The Anthropocene Review, 3*(1), 651–674.

Hewlett, S. A., et al. (2008). *The Athena factor: Reversing the brain drain in science, engineering, and technology.* Harvard Business Review Research Report, 10094.

Hill, C., Corbett, C., & St Rose, A. (2010). *Why so few? Women in science, technology, engineering, and mathematics.* Washington, DC: American Association of University Women.

Houser, H. (2016). *Climate visualizations as cultural objects. Teaching climate change in the humanities.* New York: Routledge.

Howarth, R., et al. (2012). Nitrogen fluxes from the landscape are controlled by net anthropogenic nitrogen inputs and by climate. *Frontiers in Ecology and the Environment, 10*(1), 37–43.

Jacob, D., et al. (2007). An inter-comparison of regional climate models for Europe: Model performance in present-day climate. *Climatic Change, 81*(1), 31–52.

Johnson, M. (2006). The tide reversed: Prospects and potentials for a postcolonial archaeology of Europe. In*Historical archaeology* (pp. 313–331). Oxford: Blackwell Publishing.

Jones, A., et al. (2010). Geoengineering by stratospheric SO_2 injection: Results from the Met Office HadGEM2 climate model and comparison with the Goddard Institute for Space Studies Model. *Atmospheric Chemistry and Physics, 10*, 5999–6006.

Kaplan, J. O., Krumhardt, K. M., & Zimmermann, N. (2009). The prehistoric and preindustrial deforestation of Europe. *Quaternary Science Reviews, 28*(27), 3016–3034.

Klinghoffer, A. J. (2006). *The power of projections: How maps reflect global politics and history.* Westport: Greenwood Publishing Group.

Knutti, R., & Sedláček, J. (2014). Robustness and uncertainties in the new CMIP5 climate model projections. *Nature Climate Change, 3*(4), 369–373.

Kravitz, B., et al. (2013). Climate model response from the geoengineering model intercomparison project (GeoMIP). *Journal of Geophysical Research-Atmospheres, 118*, 8320–8332.

Kress, G. R., & Van, L. T. (1996). *Reading images: The grammar of visual design.* Psychology Press.

Kuhn, T. S. (1977). Objectivity, value judgment, and theory choice. In*Arguing About Science* (pp. 74–86). New York: Routledge.

Lacey, H. (1997). The constitutive value of science. *Principia, 1*(1), 3–40.

Lee, C. P. (2007). Boundary negotiating artifacts: Unbinding the routine of boundary objects and embracing chaos in collaborative work. *Computer Supported Cooperative Work, 16*, 307–339.

Lenton, T. M., & Vaughan, N. E. (2009). The radiative forcing potential of different climate geoengineering options. *Atmospheric Chemistry and Physics, 9*(15), 5539–5561.

Llanillo, P., Jones, P. D., & Von Glasow, R. (2010). The influence of stratospheric sulphate aerosol deployment on the surface air temperature and the risk of an abrupt global warming. *Atmosphere, 1*(1), 62–84.

Lloyd, E. (1996). Pre-theoretical assumptions in evolutionary explanations of female sexuality. In E. F. Keller & H. E. Longino (Eds.), *Feminism and science* (pp. 91–102). Oxford: Oxford University Press.

Longino, H. E. (1987). Can there be a feminist science? *Hypatia, 2*(3), 51–64.

Longino, H. E. (1990). *Science as social knowledge: Values and objectivity in scientific inquiry.* Princeton: Princeton University Press.

Longino, H. (1993). Subjects, power, and knowledge: Description and prescription in feminist philosophies of science. In L. Alcoff & E. Potter (Eds.), *Feminist Epistemologies* (pp. 101–120). New York: Routledge.

Longino, H. (1995). Gender, politics and the theoretical virtues. *Synthese, 104*(3), 383–397.

Longino, H. E. (1996). Cognitive and non-cognitive values in science: Rethinking the dichotomy. In L. H. Nelson & J. Nelson (Eds.), *Feminism, science, and the philosophy of science* (pp. 39–58). Dordrecht: Kluwer Academic.

Longino, H. E. (1997). Cognitive and non-cognitive values in science: Rethinking the dichotomy. In L. H. Nelson & J. Nelson (Eds.), *Feminism, science, and the philosophy of science* (pp. 39–58). Boston: Kluwer.

Longino, H. E. (2001). What do we measure when we measure aggression? *Studies in History and Philosophy of Science Part A, 32*(4), 685–704.

Longino, H. (2002). Reply to Philip Kitcher. *Philosophy of Science, 69*(4), 573–577.

Longino, H. E. (2004). How values can be good for science. In P. K. Machamer & G. Wolters (Eds.), *Science, values, and objectivity.* Pittsburgh: University of Pittsburgh Press.

Longino, H. E. (2005). Can there be a feminist science? In A. E. Cudd & R. O. Andreasen (Eds.), *Feminist theory: A philosophical anthology* (pp. 210–217). Oxford/Malden: Blackwell Publishing.

Longino, H. E. (2006). Theoretical pluralism and the scientific study of behavior. *Scientific Pluralism, 19*, 102–131.

Longino, H. E. (2008). Values, heuristics, and the politics of knowledge. In M. Carrier, D. Howard, & J. Kourany (Eds.), *The challenge of the social and the pressure of practice, science and values revisited* (pp. 68–86). Pittsburg: University of Pittsburg Press.

Longino, H. E. (2013). *Studying human behavior: How scientists investigate aggression and sexuality*. Chicago: University of Chicago Press.

Longino, H. E., & Lennon, K. (1997). Feminist epistemology as local epistemology. *Proceedings of the Aristotelian Society, Supplementary Volumes, 71*, 1–35.

Longino, H. E., et al (Eds.). (2006). *Scientific pluralism* [electronic resource] (Vol. 19). Minnesota: University of Minnesota Press.

Lunt, D. J., et al. (2008). 'Sunshade world': A fully coupled GCM evaluation of the climatic impacts of geoengineering. *Geophysical Research Letters, 35*(12), L12710.

McInnes, C. R. (2010). Space-based geoengineering: Challenges and requirements. *Proceedings of the Institution of Mechanical Engineers, Part C: Journal of Mechanical Engineering Science, 224*(3), 571–580.

McMullin, E. (1983). Values in science. In P. Asquith & T. Nickles (Eds.), *PSA 1982*. East Lansing: Philosophy of Science Association.

Mearns, L. O., et al. (2001). *Climate scenario development*. IPCC. http://www.grida.no/Climate/ipcc_tar/wg1/pdf/TAR-13.pdf. Accessed 14 Nov 2016.

Morton, O. (2015). *The planet remade: How geoengineering could change the world*. Princeton: Princeton University Press.

Oakley, A. (1998). Gender, methodology and people's ways of knowing: Some problems with feminism and the paradigm debate in social science. *Sociology, 32*(4), 707–731.

Olofsson, J., & Hickler, T. (2008). Effects of human land-use on the global carbon cycle during the last 6,000 years. *Vegetation History and Archaeobotany, 17*(5), 605–615.

Pawson, S., et al. (2008). Goddard Earth observing system chemistry-climate model simulations of stratospheric ozone-temperature coupling between 1950 and 2005. *Journal of Geophysical Research: Atmospheres, 113*, D12.

Pittock, A. B. (1993). Climate scenario development. In A. J. Jakeman, M. B. Beck, & M. J. McAleer (Eds.), *Modelling change in environmental systems* (pp. 481–503). Chichester/New York: Wiley.

Potochnik, A. (2012). Feminist implications of model-based science. *Studies in History and Philosophy of Science Part A, 43*(2), 383–389.

Preston, C. J. (2013). Ethics and geoengineering: Reviewing the moral issues raised by solar radiation management and carbon dioxide removal. *Wiley Interdisciplinary Reviews: Climate Change, 4*(1), 23–37.

Randall, D. A., et al. (2007). Climate models and their evaluation. In S. D. Solomon et al. (Eds.), *Climate change 2007: The physical science basis. Contribution of Working Group I to the Fourth Assessment Report of the IPCC (FAR)* (pp. 589–662). Cambridge: Cambridge University Press.

Rasch, P. J, Tilmes, S, Turco, R. P, Robock, A Oman, L, Chen, C, Stenchikov, G. L, & Garcia, R. R. (2008, November). An overview of geoengineering of climate using stratospheric sulphate aerosols. *Philosophical Transactions. Series A Mathematical, Physical, and Engineering Sciences, 366*(1882), 4007–4037.

Robock, A. (2014a). Geoengineering the climate system. In R. E. Hester & R. M. Harrison (Eds.), *Issues in environmental science and technology* (pp. 162–185). Cambridge: Royal Society of Chemistry.

Robock, A. (2014b). Stratospheric aerosol geoengineering. In R. E. Hester & R. M. Harrison (Eds.), *Geoengineering of the climate* (pp. 162–185). Cambridge: The Royal Society of Chemistry.

Robock, A., Oman, L., & Stenchikov, G. L. (2008). Regional climate responses to geoengineering with tropical and Arctic SO_2 injections. *Journal of Geophysical Research: Atmospheres, 113*, D1601–D16101. https://doi.org/10.1029/2008JD010050.

Robock, A., MacMartin, D. G., Duren, R., & Christensen, M. W. (2013). Studying geoengineering with natural and anthropogenic analogs. *Climatic Change, 121*(3), 445–458.

Rueschemeyer, D., Stephens, E., & Stephens, J. (1992). *Capitalist development and democracy*. Chicago: University of Chicago Press.

Sarewitz, D. (2011). The voice of science: Let's agree to disagree. *Nature, 478*(7), 2011.

Schneider, B. (2012). Climate model simulation visualization from a visual studies perspective. *Wiley Interdisciplinary Reviews: Climate Change, 3*(2), 185–193.

Seidman, S. (1983). Modernity, meaning, and cultural pessimism in Max Weber. *Sociology of Religion, 44*(4), 267–278.

Seinfeld, J. H., & Spyros, N. P. (2016). *Atmospheric chemistry and physics: From air pollution to climate change*. Hoboken: Wiley.

Shackley, S. (2001). Epistemic lifestyles in climate change modeling. Changing the atmosphere: Expert knowledge and environmental governance. In C. A. Miller & P. N. Edwards (Eds.), *Changing the atmosphere: Expert knowledge and environmental governance* (pp. 107–133). Cambridge: MIT Press.

Shepherd, J. G. (2009). *Geoengineering the climate: Science, governance and uncertainty*. Royal Society: London.

Shohat, E., & Stam, R. (Eds.). (1994). *Unthinking Eurocentrism: Multiculturalism and the media*. New York: Routledge.

Smith, N. (2010). *Uneven development: Nature, capital, and the production of space*. Atlanta: University of Georgia Press.

Soden, B. J., Wetherald, R. T., Stenchikov, G. T., & Robock, A. (2002). Global cooling after the eruption of Mount Pinatubo: A test of climate feedback by water vapor. *Science, 296*, 727–730.

Star, S. L., & Griesemer, J. R. (1989). Institutional ecology,translations'and boundary objects: Amateurs and professionals in Berkeley's Museum of Vertebrate Zoology, 1907–39. *Social Studies of Science, 19*(3), 387–420.

Stainforth, D. A., et al. (2007). Confidence, uncertainty and decision-support relevance in climate predictions. *Philosophical Transactions of the Royal Society of London A: Mathematical, Physical and Engineering Sciences, 365*(1857), 2145–2161.

Stone, N. J., & English, A. J. (1998). Task type, posters, and workspace color on mood, satisfaction, and performance. *Journal of Environmental Psychology, 18*(2), 175–185.

Tellis, W. (1997). Application of a case study methodology. *The Qualitative Report, 3*(3), 1–17. www.nova.edu/ssss/QR/QR3-3/tellis2.html. Accessed 14 Oct 2016.

Trenberth, K. E., & Dai, A. (2007). Effects of Mount Pinatubo volcanic eruption on the hydrological cycle as an analog of geoengineering. *Geophysical Research Letters, 34*, 15. http://onlinelibrary.wiley.com/doi/10.1029/2007GL030524/full. Accessed 1 Nov 2016.

Tuana, N. (Ed.). (1989). *Feminism and science*. Bloomington: Indiana University Press.

University of Exeter. (2008). Humans implicated in prehistoric animal extinctions with new evidence. *ScienceDaily*. www.sciencedaily.com/releases/2008/08/080811200028.htm. Accessed 1 Nov 2016.

Wajcman, J. (1991). *Feminism confronts technology*. Pennsylvania: Penn State Press.

Williams, J. W., Stephen, T., & Jackson, S. T. (2007). Novel climates, no-analog communities, and ecological surprises. *Frontiers in Ecology and the Environment, 5*(9), 475–482.

Woodward, J. (2003). *Making things happen: A theory of causal explanation*. Oxford: Oxford University Press.

Wylie, A. (1995). Doing philosophy as a feminist: Longino on the search for a feminist philosophy. *Philosophical Topics, 23*(2), 345–358.

Chapter 4
Ontological Heterogeneity

Abstract In this Chapter, the feminist virtue of ontological heterogeneity is addressed as it relates to SO_2 climate engineering. In particular, the scientific practices that underlie sulfate geoengineering studies with respect to whether it fulfills the principles of feminist epistemology laid out by Helen Longino and FCE is examined. Overall, it is argued that this approach to climate change mitigation does not.

Keywords Heterogeneity · Modeling · Feminist empiricism · Helen Longino · Geoengineering · Sulphate · Volcanoes · Rural · Urban

Longino's supplementary virtues work to augment empirical adequacy and simultaneously explain and assess the decisions made with respect to theory choice and models as well as articulating postulates to guide feminist science. Fundamentally, their feminist bona fides are made manifest by making gender a central part of the inquiry and preventing its disappearance as "a bottom line requirement of feminist knowers" (Longino 1994, 481). These virtues also thematize the central concern of feminist epistemology which is aimed at "understanding the systemic nature of our epistemic practices and assessing these practices accordingly" with particular attention paid to distortions that arise from, among other things, gendered power relations. Heidi Grasswick calls this "knowing well" (Grasswick 2014, 222–223).

In this Chapter, the feminist virtue of ontological heterogeneity is addressed as it relates to SO_2 climate engineering. In particular, I examine the scientific practices that underlie sulfate geoengineering studies with respect to whether they fulfill the principles of feminist epistemology laid out by Helen Longino and FCE. Cumulatively, and through a wide-ranging analysis, I argue that this approach to climate change mitigation, in fact, does not.

T. Sikka, *Climate Technology, Gender, and Justice*, SpringerBriefs in Sociology, https://doi.org/10.1007/978-3-030-01147-5_4

Ontological Heterogeneity

The second significant feminist scientific virtue of note is that of ontological hetero-geneity. As stated, this virtue doubles down on the need to thematize specificities, differences and particulars over abstractions, standardizations, and generalizations as they relate to scientific practice. Because standardization and simplicity are tra-ditionally given ontological priority, Longino argues that what is required is a shift in thinking such that monocausality, i.e. the tendency to, "treat apparently different entities as versions of a standard or paradigmatic member of the domain," and wherein "differences are regarded as eliminable" (Longino 1995, 378), are chal-lenged. Rather, differences ought be seen as a resource and given equivalent consid-eration and not as a simplistic or inferior deviation from a norm or type. Longino points out that this 'theory of inferiority' stems from an ontological prioritization of the straight, middle class, white male and their perspectives. A useful example she gives of such a predisposition has to do with the divergence between how scientists traditionally describe the process of fertilization, in which the aggressive and active sperm is seen to fertilize a passive egg. Instead of a model in which fertilization is seen as an interactive and cooperative process as in the model of gamete fusion (Longino and Lennon 1997). In this case, gender ideology inserts itself into how fertilization occurs despite "evidence of the egg's activity in binding and drawing the sperm as well as blocking out extra sperm and the evidence for the very weak propulsion of sperm tails" (Spanier 1995, 24).

In the very distinct case of solar geoengineering, these kinds of explicitly gen-dered divisions are not readily apparent. Yet, by drawing attention to the social effects of ontological monism and the prioritizing of simplicity, a feminist endorse-ment of theories that resist "unicausal accounts of development in favor of accounts in which quite different factors play causal roles" (Longino 2010, 22) becomes clear.

This Chapter begins by establishing how solar climate engineering research engages in the kind of ontological monism Longino warns against by examining selected studies including the GeoMIP research discussed in the previous Chapter. In addition to ontological monism, I demonstrate how sulphate geoengineering studies tend to collapse different entities into variations of one standard and concep-tualize differences as lacking consequence – which further indications that onto-logical heterogeneity is lacking. Also considered is how one might move away from approaches to science that grant privilege to theories that purport a sole cause and, in doing so, reflect on alternatives to scientific practices that treat "apparently differ-ent kinds of entities as merely different versions of a single standard kind of entity," and that address those "differences as eliminable through decomposition of entities into a single basic kind" (Kourany 2003, 6). Yet, it should be noted that not all sul-phate SRM studies are uniformly mono-causal. Some do show a partiality for a broadening of the discursive and experimental space, but not to the degree that would warrant the marque of ontological heterogeneity.

For example, a new simulation model developed in 2017 by researchers from the National Centre for Atmospheric Research (NCAR), Pacific Northwest National Laboratory (PNNL), and Cornell University is capable of adjusting to account for different levels of sulphur dioxide injections as well as location-specific consequences. Approaches of this sort would more aptly account for ozone damage as well as altered precipitation levels in manner that is better able to explicate unforeseen side-effects on marginalized groups as well as focusing attention on multiple outcomes, variability, and localness (NCAR/UCAR 2017). What this tells us about the evolution of these debates is that there room for change in the structure of scientific practice consistent with the principles articulated by Longino as well as changing values, discourses, policies, and objectives.

Another example can be found in Kravitz et al.'s study which, in addition to the factors of temperature and radiation, also discusses the effects of geoengineering on: sea ice levels – which has indeterminate results; hydrology, with precipitation increasing for the tropics; and terrestrial net productivity, in which the health of the biosphere would generally benefit (Kravitz et al. 2013). While this expansion is laudable, the problem is that the scope of the questioning and openness to genuine alternatives is still lacking. Longino is clear that for this to occur, we need to not only describe science in context, wherein novel factors such as geographical time, space, and place, historical context, power relations, gender ideology, and other taken for granted assumptions and norms are unpacked, but also to develop a strong normative theory of science that maintains that "the structures of cognitive theory themselves must change" (Longino 1996, 278).

Yet it could, and has, been argued that these criticisms of mainstream science are unwarranted as these studies are designed specifically to the test, model, and analyze a very particular set of questions. Historically, scientific methodology aims for instrumental reliability based on observation and displays a "preference for theories having the property or properties…[of] simplicity or parsimony" or what "scientists often call elegant (or elegance) instead" (Boyd, 349, 1991). Consequently, opening up the scope of analysis to new factors, dynamics, and subjects might risk diluting this central objective. However, as Longino makes clear, this preference for simplicity fails to recognize "difference [as] a resource, not failure" (Longino, 1, 1994, 477) and overstates the extent to which heterogeneity risks diluting scientific focus.

In regards to whether solar geoengineering studies tend towards theories that purport a sole cause of particular phenomena, it is the case that the majority of scientists working in this area focus on the effects of carbon dioxide at the expense of other potential causes of climate disruption. This particular issue has been discussed at length in Chapter one, but it bears repeating that most sulphate geoengineering studies that examine its potential efficacy in mitigating average temperature increases do so by ignoring causes of warming beyond CO_2. In a cursory look at the most cited studies of SO_2 geoengineering, it is CO_2 mitigation that is the objective of the entire enterprise (Ricke et al. 2012, Irvine et al. 2010; Crutzen 2006; Izrael et al. 2009). In the Kravitz et al. study, it is only CO_2 that is built into the model despite the phrase 'greenhouse gas emissions' being used a handful of times. This is illustrative of a larger tendency of insisting on one standard marker of climate change at the expense of others, e.g. chlorofluorocarbons and nitrous oxide, which weakens ontological heterogeneity.

Overall, it is the radiative forcing caused by increased CO_2 specifically which this particular geoengineering approach aims to address. As Victor et al. conclude in their comprehensive study of SO_2 climate engineering, "one kilogram of well placed sulfur in the stratosphere would roughly offset the warming effect of several hundred thousand kilograms of carbon dioxide" (Victor et al. 2009). While it is true that, the objective of these studies is to demonstrate the ability of sulphate aerosols to mitigate warming full stop, as stated, this mono-causality overlooks other causes of climate change including methane emissions, nitrous oxide, fluorinated gases, deforestation, ozone depletion, burning of fossil fuels, changes in land use, and population increase. For Longino, this limits diversity in how an object or process is studied. It also potentially limits creative thinking with respect to how, for example, the 16% of GHG emissions from methane could be at least partially mitigated through efficiencies in pasture and grazing, the trading of methane credits, and manipulating animal diets (Gerber et al. 2013; Lascano and Cardenas 2010). This is particularly important since it is estimated that, as a result of higher incomes and general population growth, methane emission are likely to increase further (Key and Tallard 2012).

Another issue is related to causality and the need to open up to heterogenous factors that has to do with the subject of causal responsibility. Responsibility is an essential value and in feminist epistemology it is understood as differential, contextual, variegated and distributed. Responsibility in science, however, tends to be expressed as mono-, or in the best case bi-, causally allocated. Moreover, it is also the case that the majority of studies of SO_2 geoengineering that aim to defend its use, responsibility for causing climate change is evenly distributed through a process of implicit absences – e.g. by not mentioning any individual countries, industries, or behaviors – coupled with the mobilization of the royal 'we' which further de-individualizes findings and recommendations. For example, in the Kravitz et al. study, there is no mention of discrete countries or specific regions that might indicate responsibility associated with groups of countries, although some hemispheric differentiations are made (see below). There are, however, several uses of the words "global," "globally," "global average," "global difference," and "global carbon cycle." Consequently, the world is treated as a cosmopolitan whole which is not only unhelpfully homogenizing on empirical grounds, but also serves to render responsibility and decision-making similarly diffuse. Saurin (1996) argues that it is scientific research itself that has constructed this globalizing worldview which works to set out a practicable but simplistic frame for how to discuss climate change and its solutions. As such, it becomes par for the course that when apportioning responsibility and isolating causality, considerable complexity overlooked.

For instance, is demonstrably true that, distributionally, the countries that emit the largest proportion of greenhouse gases are those with the large industrial bases, agricultural industries, and highest standards of living (the latter of which demand consistent heating, electricity, transportation (cars), and food (particularly meat)).

China, at 28%, the US, at 16%, the EU, at 10%, and India, at 6%, round out the top four largest emitters globally (Boden et al. 2010). However, these numbers look much different on an emission per capita basis in which Canada, the US, the Russian Federation and Japan take the top spots (Ge et al. 2014). With respect to intensity, it is Indonesia, China, the Russian Federation and Canada whose emissions are rising most rapidly and, vis-à-vis the countries most answerable for cumulative emissions, the US, EU, and China carry the largest responsibility (Ge et al. 2014).

Once again, the problem here is that by focusing on solar geoengineering as a potential solution to a global problem (which, like the GeoMIP piece, most studies do) (Dickinson 1996; Niemeier et al. 2013; Keith et al. 2010)), and by coupling this with an emphasis on universalizing solutions, the result is a narrowing of complexity and causation, and, consequently, an inability to engage in a science that can fruitfully harness pluralism, diversity, and heterogeneity. Longino is clear that when science goes global, a whole host of complications are made manifest. She asks, under these conditions, "How can individual societies or communities maintain the control envisaged? And what happens when different equally advantaged or even differently advantaged societies embrace values and agendas that will conflict when put into action?" (Longino 2013a, b, 568). These questions are particularly salient with respect to climate engineering.

Studies of SO_2 climate engineering that focus on global carbon emissions, wherein the world and humanity *at large* are seen as *the* driving factors, do not account for the fact that differences also exist with respect to economic sectors. According to the US Environmental Protection Agency (EPA), by economic sector, global greenhouse gas emissions break down with electricity and heat production (coal, oil, natural gas) taking the lead at 25%; industries that rely heavily on fossil fuels for energy at 21%; agriculture, forestry and land use, including crops, livestock and deforestation, is responsible for 24%; transportation, e.g. gas and diesel, 14%; buildings, including the burning of fuel for cooking and heating, 6%; and other energy uses, e.g. the extraction, refining, transport of fuel, 10% (EPA 2016).

It is important, if we are to encourage a pluralist and heterogenous practice of science, to expand the kinds factors examined such that, in the case of sourcing CO_2 emissions, questions around what specific sectors that bear inordinate responsibility for climate disruption can do to cut emissions, perhaps even to make geoengineering unnecessary, are emphasized. Taking on emissions as a global phenomenon ignores this. On the level of science, a more pluralist approach should include a discussion of how these different sources of emissions are likely to be affected by sulphate geoengineering and how to calibrate for different kinds of emissions that are associated with each sector. For example, it is the case that different sectors emit different levels CO_2 in concert with other GHGs that can be more or less disruptive. According to a book written for the UN Food and Agricultural Organization (FAO), livestock produce over 100 other polluting gases including carbon dioxide, consume inordinate amounts of water, and drive deforestation through overgrazing (Steinfeld et al. 2006). This provides us with another perspective.

Additionally, differentiation and multi-causality of this kind has its impacts where climate disruptions are more likely to take place and the spaces in which sulphate engineering might be more or less useful and/or harmful. Accordingly, the influences of both are also going to be unequally felt between countries depending on where the sites of GHG production, use, and potential deployment of climate engineering are situated. Understanding this multiplies causation in ways that are left out of the most cited geoengineering studies which tend to envision the world as a monolithic whole that moves towards "a single modern pattern" using a problematic and "synthesizing macro-sociological perspective" as demonstrated further below (Khondker and Schuerkens 2014).

Yet, it should be acknowledged that not all climate engineering studies engage in unmitigated wholism. Alan Robock et al. for example, in an article for *Atmospheres*, focus specifically on the regional impacts of SO_2 geoengineering by paying attention to the extent to which "Both tropical and Arctic SO_2 injection would disrupt the Asian and African summer monsoons, reducing precipitation to the food supply for billions of people" (Robock et al. 2008, D16101). Kravitz et al., in their piece, also regionalize their findings by referring to the Tropics, the Arctic, and the Antarctic. Yet neither are remotely granular enough to meet Longino's feminist "bedrock commitment to a nonreductionist, dialectical" (Longino 2008, 78) view of science. In another piece, Forester et al. goes through several alternative readings of SRM science based on differences in the modeling of aerosol properties, cloud physics, and interactions between clouds and aerosols across models (Forster et al. 2007), as well as challenging consensus around the potential increases in plant productivity. Yet even in critical pieces that question scientific consensus, a heterogenous approach is still lacking. A truly heterogenous science should be committed to avoiding a "reliance on monist assumptions in interpretation or evaluation coupled with an openness to the ineliminability of multiplicity in some scientific contexts" (Kellert et al. 2006, xiii). For this to be achieved in contemporary geoengineering research, much more work needs to be done.

This is where an explicitly gendered analysis becomes helpful since it is also the case that the causal effects of climate change, as well as the potential repercussions of SO_2 climate engineering, are gendered. For ontological heterogeneity to play its part in feminist science, issues of gender must be made manifest – particularly with respect to causality and effect. In the case of SO_2 climate engineering, a straight line from causality to effect can be drawn on the level of which parties bear the most responsibility for climate change to begin with. Ecofeminist scholars working on this have pointed to climate change as an outcome of an industrial and militaristic approach to development supported by patriarchy and colonialism (Hessing 1993; Nhanenge 2011). According to Sherilyn MacGregor, a noted ecofeminist, "A feminist response is to point out that by 'scientizing' and 'securitizing' it, climate change is constructed as a problem that requires the kinds of solutions that are the traditional domain of men and hegemonic masculinity" (MacGregor 2009, 128). A very similar argument can be made about climate engineering – i.e. that its very ethos serves to simply extend an impoverished view of the world based on masculine values.

However, the connection of particular significance in the context of FCE lies with the material effects of sulphate geongineering on marginalized communities and especially women (Dankleman 2002). This is a particularly vexing subject in that it draws attention to the need to bring a local and multiscalar mode of analysis into geoengineering research without re-marginalizing, reifying, or homogenizing women as a category. Currently, there are a notable number of publications documenting the ways in which women are unevenly impacted by climate change whose findings, observations, and estimations can be extrapolated to SO_2 climate engineering (Nelson et al. 2002; Dankelman 2010; Nath and Behera 2011). Research from the United Nations Development Programme (UNDP), for example, details how women in the developing world, as a result of climate disruptions, suffer disproportionate difficulties in safeguarding consistent access to fuel, food, water, and healthcare. This is in addition to having to withstand differentiated access to resources including "assets and credit and [unequal] treatment by markets and formal institutions (including the legal and regulatory framework)" which "constrain women's opportunities" (Habtezion 2013). Unequal power to influence policy and decision-making, in addition to socio-cultural factors, multiplies this effect. As Nelson et al. contends, "vulnerability to environmental degradation and natural hazards [are] articulated along social, poverty and gender lines" in a manner that "can increase both women's workload and their vulnerability, as their access to already scarce resources decreases" (Nelson et al. 2002, 51–52).

However, it is also the case that this narrative can feed into a paradoxical logic wherein women are framed as both as victim and saviour. Seema Arora-Johnsson (2011) critiques the victim narrative by pointing out that research on the feminization of poverty, the real effects of more access to land, and the mortality rates of women in natural calamities leave much to be desired. She also argues that the tendency to use the victim logic to frame women as virtuous – wherein they are then given the task of solving climate change – further burdens them with additional caring work. As such, it is important, in this context, to avoid the tendency to validate the women as victim trope, which can be depoliticizing, while simultaneously continuing to champion "projects and programmes that sidestep existing disproportionate workloads and gendered hierarchies" (Resurrección 2013, 34). These divisions and burdens are likely to be felt by women in the course of climate engineering as well – particularly with respect to the likelihood of increased droughts and monsoons due to disrupted precipitation levels that will lead to problems in accessing water with direct and indirect impacts on biodiversity and food security (Scheffran et al. 2016; Baum et al. 2015). Thus, any program of mitigation must consider how "vulnerabilities are…shaped by norms of sex and gender" as well attending to the fact that "ecological and social vulnerabilities should inform harm-reduction strategies, and as resources are directed toward communities [e.g. women] facing imminent threats, claims about vulnerabilities will become increasingly influential" (Cuomo 2011, 694).

Problems can also be seen in how the urban/rural and producer/consumer binaries are used. The former is manifest in scientific geoengineering research primarily through a conspicuous absence or by elevating the urban over the rural. It should be noted that there is a significant amount of tension and contradiction inherent in in the urban/rural binary rooted in the historical and value laden assumption that cities are "necessarily products of economic modernisation and are characterised by forms of citizenship, bureaucratic management, infrastructure and services, and cosmopolitanism familiar to Anglo- Europeans" (Angelo 2017, 2; Roy 2009).

Pre and post-Cold War development discourse mirrors this predilection for models of development that prioritize the urban over the rural by academics, politicians and corporations. Developmentalism, as a model of international development, was itself manifest in a policy-driven privileging of Western and industrial modes of capitalist growth and development for which cities were seen as signifiers of progress, cosmopolitanism, and modernity (Wallerstein 1992; Robinson 2011, Sheppard et al. 2013).

It is notable that, contemporaneously, it is the city and other urban environments, rather than rural ones, that are increasingly seen as climate saviors with respect to the growing use of mass transit, energy efficiency, the so-called sharing economy (despite its myriad contradictions), innovative urban design, and urban agriculture which largely conflicts with the belief that climate engineering is immediately necessary (Leichenko 2011; Broto and Bulkeley 2013).

A salient example of this can be found in the fact that the observations on which stratospheric sulphate geoengineering extrapolations and models are made tend to be based on evidence taken from volcanic monitoring stations located primarily in urban areas. As such, variations in dimming in non-urban rural, forest, tropical and less populated areas are left unexplored (Alpert et al. 2005; Schwartz 2005). A lack of gathered evidence from these regions leads to a limitation in model-associated inputs, which then places further restrictions on the scope of knowledge and associated conclusions. Making the case for the probable success of sulphate geoengineering with limited inputs, stratified along urban/rural lines, does not encourage ontological heterogeneity. It also feeds into a tradition of sidelining rural issues for more urban ones despite the fact that there are more people living in rural than urban communities.

It is also the case that rural and/or remote areas of the world are likely to be differentially affected by the impacts of climate geoengineering. While the Kravitz et al. article does point out that tropical regions are likely get less rainfall, little beyond this fact is discussed. A significant corollary of this outcome is examined by Sedjo (2010) who argues that a lack of rainfall in Amazonian areas, as a result of stratospheric geoengineering, will likely lead to larger and more intense fires – costing lives (human and animal) and money for reconstruction and environmental mitigation. It is also probable that, these disruptions will interrupt consistent supplies of clean water and food. Differentiated global cooling can decrease agricultural productivity leading to the possibility of distress and as well as unrest (Brewer 2007; Trenberth and Dai 2007). This follows a long line of evidence based argumentation showing that it is vulnerable and largely rural communities that are suffering most from the consequences of climate change – from atoll-countries like Kirabati and Micronesia who are risk of sea level rise, to Inuit communities in the Arctic that are suffering from retreating snow and ice, and countries like Bangladesh which have become more and more prone to flooding (Chen et al. 2013; Warrick and Ahmad 2012).

Another way of thinking about the urban/rural binary, as it intersects with the consumer/producer divide, is as it relates to country to country relations and North/ South divisions whose present status are shaped by colonial histories that continue to be founded on carbon intensive resource exploitation. While this subject has been touched upon in relation to contemporary nation-to-nation relations, it is also important to examine this with respect to the historical consequences of colonization and the structural inequalities that continue to exist as a result. As Roberts and Parks argue, "The issue of global climate change – which itself is characterized by tremendous inequalities in vulnerability, responsibility, and mitigation – can therefore not be viewed, analyzed, or responded to in isolation from the larger crisis of global [and historical] inequality" (Roberts and Parks 2006, 27). This inequality in "vulnerability, responsibility, and mitigation" is why the countries of the global South continue to argue that the bulk of their historical greenhouse gas emissions are not only small in relation to their populations and in comparison to those more powerful and developed nations, but also that the carbon they did and continue to emit is in the service of countries, markets, and citizens of the global North (Gardiner 2004; Müller et al. 2009).

Moreover, it is also argued that it is unfair for rich countries, who have benefited materially from carbon intensive development, to deny countries of the global South these same advantages. The issue of differentiated responsibility for future emissions cuts is particularly important in this context (Baer et al. 2009; Neumayer 2000). As stated, in regards to geoengineering, these concerns demonstrate the necessity for granular and heterogenous models of thought that should be incorporated into 'scientific' geoengineering research which, as has been established, is always value laden. A decision has been made that issues around responsibility, urban/rural and producer/consumer differences, and historical power relations are not evocative in the context of hard science. The question that arises is how these interests and concerns can be incorporated into contemporary geoengineering research and why, at this time, they are not.

Accordingly, a more thoughtful thematizing of these distinctions is important, as is a complicated mediation between global and local issues and effects, and conversations about the historical marginalizing of rural communities and poorer regions and peoples as it relates to how testing of geoengineering might unfold – i.e. where and on whom? Morrow et al. (2009) take this up by drawing a troubling parallel between nuclear weapons testing, which in the US took place near poor, rural communities, the subjecting of the Black populations to medical experimentation; and prospective geoengineering tests which risks following the same pattern. Socially ahistorical thinking, in relation to the urban/rural binary, is symptomatic of a significant portion of sulphate geoengineering research which perpetuate these omissions.

Accordingly, it is important to reiterate that calls for heterogeneity along these lines do not lead to an unproductive expansion of variables such that larger issues, e.g. the overall efficacy of solar geoengineering, gets lost in specificities. Rather, its objective is to trade simplicity for complexity in order to reveal gender relations and other ethico-political concerns. Nor does this approach invite relativism since ontological heterogeneity, in line with feminist epistemology, serves to harness the strength of difference into "crafting objectivity" as "heterogenous networking – tying as many things together as possible" (Hodder 1997, 10).

Finally, when this urban/rural binary is looked at more closely, further insights can be made as it relates to conventional understandings of responsibility which, in turn, opens up alternative avenues for mitigation and abatement. In particular, a more scalar approach can draw attention to the fact that causality can also be divided along the axis of cities, which are often thought to require high levels of energy use to sustain the kinds of lifestyles found urban areas, and rural communities, which are believed to have much the lower emissions due to lower incomes and more sustainable consumption habits. This thesis, however, is made more complicated when one realizes that the bulk of "agriculture, deforestation and emission from heavy industries, fossil- fuelled power stations and high-consumption households...[in affluent countries] are not located in cities" proper and that wealthy individuals that consume high levels of energy live in rural areas as well" (Satterthwaite 2008, 539; Dodman 2009).

In fact, in developing countries like India, increases in consumption in rural areas have actually outpaced growth in urban ones (NSSO 2011). Additionally, in developed countries the urban/rural distinction in consumption is further problematized when one realizes that, on a private level, "only a diminishing share of the earnings is consumed as the level of earnings grow. This has an equaling effect on the carbon consumption of different lifestyles" (Heinonen and Junnila 2011, 1242).

While it is the case that the sites of consumption for which the bulk of energy intensive activities occurs is not always in urban centers, what is most important to recognize is that drawing cut and dry distinctions, or engaging in generalizations for the sake of simplicity, undermines alternative views. Science-based geoengineering research does not grapple with these subjects sufficiently. Questioning the globalizing and urban/rural narratives, with an eye to heterogeneity, opens up space from which to consider novel forms of abatement and mitigation that could challenge the need for geoengineering at all. On the level of urban mitigation, this includes efficiencies in heat and electricity, new energy sources, better building schemes, public transport, fewer consumer goods, and less travel, while also leaving room for cuts in absolute levels of consumption. In rural communities, it would require the incorporation of conservation in agriculture (in relation to tillage and water use), the move towards wind and solar energy, curtailment of emissions from manufacturing plants, the expansion of forests, and the phasing out of coal use (Fink 2013). The preponderance of scientific studies of sulphate climate engineering fails to examine these distinctions.

Turning back to the producer/consumer relation, this binary aims to answer the question of whether it is the consumer whose demands for goods and services (and consumption of both) that portends higher emissions or the producers of said goods and services. This then leads to several pointed quesitons about the motivations, choice, responsibility, and interests associated with sulphate climate engineering as well. Briefly, since these categories overlap significantly with global/local binary articulated above, it is important to note how differentiated, diffuse, and therefore difficult sourcing greenhouse gas emissions, and therefore placing responsibility for action, can be. For example: What are the differences between emissions emanating from oil, gas, and coal production versus those used by consumers? Who bears the greater responsibility – the consumers who consume or the producers who produce for those consumers? And how would geoengineering intersect with these interests – particularly when solar geoengineering is implicated in the political economy of energy production as well as the politics of climate change? While climate change research has integrated sourcing into their models, with some attention paid to the producer/consumer divide, very little has been integrated into SRM geoengineering models on these subjects (Davis et al. 2011; Wyckoff and Roop 1994).

According to recent research from the journal *Climatic Change* (Heede 2014), and reported on by such publications and *The Guardian* (2013), 63% of global methane and CO_2 emissions can be traced to 90 companies engaging, largely, in the production of coal, gas and oil. Big players such as Exxon, BP, Chevron Texaco, BHP Billiton, Gazprom, and Shell are at the forefront of this list. It is notable that in the *Climatic Change* piece, the author explicitly attaches the countries to which the companies are associated (Pemex, Mexico; Total, France; Chevron, USA etc.) which further isolates the ethic of responsibility in manner not thus far taken up by scientific, model-based studies of geoengineering. It is also significant that, on the subject of responsibility, there is sparse attention paid to the fact that geoengineering fails to deal with the core causes of climate change (overconsumption, resource extraction, generalized pollution), while attending solely to the symptoms (which is an issue I return to further on).

On the consumer side, there are salient arguments to be made that were there no demand for products, services, and transportation that damage the environment, particularly in developed and middle income countries, producers would have no market to serve. This perspective places the onus for change on individual actors rather than a system that requires constant consumerism to grow (Michael 2009). Alternatively, a consumer lifestyle approach to climate change goes about this issue in a more productive way by rejecting demand instigated causality and, instead, isolates individual and/or household consumption patterns with respect to culture, technology, beliefs, income, household size etc. as a way to consider the motivations for consumption as well as possible avenues for policy and behavioural change (Bin and Dowlatabadi 2005). Geoengineering research fails to deal with these complexities as well preferring, instead, to reify monocausal, linear science at the expense of heterogenous science. Thus, a more nuanced understanding of the motivations, values, and pressures placed on consumers is needed – particularly if it expands the remit of geoengineeering research beyond the 'likely reduction in global CO_2' objective. Again, this broadening is not aimed at apportioning blame, from which to then assign responsibility for action per se, but to draw attention to the fact that the causes of warming are multiple and therefore so are solutions to it. Studies of SO_2 climate engineering, as demonstrated, tend to overlook this heterogeneity in favour of globalizing causality in a manner that limits alternative readings of the object of study. A more diverse science would encourage the kinds of local and granular analysis between individual countries, regions, the urban/rural divide as well as the fissures opened up along the lines of gender, race and class as noted above.

Longino's conception of ontological heterogeneity also emphasizes the importance of giving due attention to deviations, anomalies and, nonconformities which would, more often than not, be explained away as insignificant or trivial. This is particularly the case since, traditionally, the standardizing of knowledge serves to obscure alternatives and present one outcome as *the* desirable outcome, finding or approach. A salient example of this proclivity can be found in Ágnes Kovács (2012) work on the difference between real and ideal gases which she argues are founded on a scale of perfection/imperfection, in which the former is seen as inferior to the latter, as well as "Platonism…[and a] disregard for embodiment and individualism" (Kovács 2012).

Feminist challenge the inferiority thesis because of its roots in a framework in which,

> …the white middle class male (or the free male citizen) as the standard grant ontological priority to that type. Difference is then treated as a departure from, a failure to fully meet, the standard, rather than simply difference. Ontological heterogeneity permits equal standing for different types, and mandates investigation of the details of such difference (Longino 1992, 337).

Overall, a feminist practice of science requires looking more closely at the issue of causality in sulphate geoengineering and opening up inquiry such that diversity is included and deviations are not seen as inferior. This entails the inclusion of voices that would object to the conclusion that SO_2 climate engineering research, through idealized models and simulations, represents what Curry et al. and Kravtiz et al. argued indicate that global radiative forcing, as a result of the increased CO_2 emissions, can potentially be compensated for and mitigated by aerosol climate geoengineering (Curry et al. 2014; Kravitz et al. 2013).

Finally, on the subject of risk, it is the case that divergent interpretations of risk are not only difficult to publicize within the realm of hard science, because of the hostility towards values in science, but, as Longino argues, tend to be presented as ancillary, secondary or mitigatible when they are discussed. For instance, in the Kravitz, et al. article, it is noted that one of the likely outcomes of sulphate geoengineering is that tropics will be cooler and poles warmer against preindustrial averages (Kravitz et al. 2013). Even when specific risks are acknowledged, setting them up against such powerful statements as "Modeling studies have shown that geoengineering using stratospheric aerosols has the potential to reduce Earth's global-mean surface temperature" (Ferraro et al. 2014) (Rasch et al. 2008; Jones et al. 2010), renders it difficult to conceive of alternative readings of risk, or expanding its remit, as anything more than secondary.

This is true despite the caveats made which, while valuable, do not rise to the level of qualitative pluralism. Moreover, qualifications, as well caveats, tend to either be subsumed into general support for climate engineering or presented as ancillary pr secondary – that is, as manageable deviations from the norm. In the GeoMIP piece this is done through a simple act of mentioning the potential consequence, risk, or lacunae in knowledge and moving on. The lack of model agreement on the effects of SO_2 geoengineering on Antarctic sea ice, the inability to adequately consider potential extreme events – because they "manifest on shorter time scales" than the monthly means used (Kravitz et al. 2013, 8329), and unforeseen changes to the hydrological cycle (e.g. monsoons) are just a few examples. Consequently, alternative interpretations of the findings of studies like that of the GeoMIP are not able to adequately address issues of risk, hazard, and threat. Nor, as demonstrated, are they capable of integrating heterogenous perspectives with respect to gender, space, and population.

Geoengineering studies ignore the implications of a lack of ontological heterogeneity at their peril since it entail overlooking significant ontological and epistemological concerns related to heterogenous factors that inform causality and that have material consequences. The effects on women, minorities, local communities, the ozone, and marine life are not ancillary, nor are the "social or political structures, ecosystem dynamics, and many other potentially important issues"(Kravitz et al. 2013, 8331), mentioned, but not discussed, in the conclusion of the Kravitz et al. piece. These concerns are not only discounted as lacking seriousness but, in their qualification, are also depoliticized. The inclusion of the following statement advances this thesis by connecting a generalized optimism about SO_2 geoengineering induced mitigation with uncertainties channelled towards a call for actual tests:

> As such, while insights into the likely response to uniform solar geoengineering can be obtained, modeling a reduction in solar irradiance cannot serve as a substitute for simulating the response to increased stratospheric aerosols or other possible approaches for reducing incoming solar radiation (Kravitz et al. 2013, 8322).

As per the ontological monistic and mono-causal approach, this is a robust example of scientific practice that is "falsely generalizing and insufficiently attentive to historical and cultural diversity" (Strickland 2012, 266) when compared to what a decisively ontologically heterogenous approach would demand. Overall, the failure of SO_2 climate engineering research to include heterogenous factors into its research renders it unable to appreciate the importance of "the complexity of poverty and disadvantage" by focusing disproportionately on "linear, apolitical solutions" (Alston 2015, 3). In the next Chapter, I move on to a discussion of the feminist scientific virtue of novelty and its role, or lack thereof, in contemporary SO_2 geoengineering research.

References

Alpert, P., et al. (2005). Global dimming or local dimming?: Effect of urbanization on sunlight availability. *Geophysical Research Letters, 32*(17), 17802.
Alston, M. (2015). *Women and climate change in Bangladesh*. New York: Routledge.
Angelo, H. (2017). From the city lens toward urbanisation as a way of seeing: Country/city binaries on an urbanising planet. *Urban Studies, 54*(1), 158–178.
Arora-Jonsson, S. (2011). Virtue and vulnerability: Discourses on women, gender and climate change. *Global Environmental Change, 2*(2), 744–751.
Baer, P., et al. (2009). Greenhouse development rights: A proposal for a fair global climate treaty. *Ethics Place and Environment, 12*(3), 267–281.

Baum, S. D., et al. (2015). Resilience to global food supply catastrophes. *Environment Systems and Decisions, 35*(2), 301–313.

Bin, S., & Dowlatabadi, H. (2005). Consumer lifestyle approach to US energy use and the related CO_2 emissions. *Energy Policy, 33*(2), 197–208.

Boden, T., et al. (2010). *Global, regional, and national fossil-Fuel CO2 emissions tech*. Carbon Dioxide Information Analysis Center, Oak Ridge National Laboratory, US Department of Energy.

Boyd, R. (1991). Observations, explanatory power, and simplicity: Toward a non-Humean account. In R. Boyd, P. Gasper, & J. D. Trout (Eds.), *The philosophy of science* (pp. 349–377). Cambridge, MA: MIT Press.

Brewer, P. G. (2007). Evaluating a technological fix for climate. *Proceedings of the National Academy of Sciences, 104*(24), 9915–9916.

Broto, V. C., & Bulkeley, H. (2013). A survey of urban climate change experiments in 100 cities. *Global Environmental Change, 23*(1), 92–102.

Chen, J. L., Wilson, C. R., & Tapley, B. D. (2013). Contribution of ice sheet and mountain glacier melt to recent sea level rise. *Nature Geoscience, 6*(7), 549–552.

Crutzen, P. (2006). Albedo enhancement by stratospheric sulfur injections: A contribution to resolve a policy dilemma? *Climate Change, 77*, 211–219.

Cuomo, C. J. (2011). Climate change, vulnerability, and responsibility. *Hypatia, 26*(4), 690–714.

Curry, C. L., et al. (2014). A multimodel examination of climate extremes in an idealized geo-engineering experiment. *Journal of Geophysical Research: Atmospheres, 119*(7), 3900–3923.

Dankelman, I. (2010). *Gender and climate change: An introduction*. New York: Routledge.

Dankleman, I. (2002). Climate change: Learning from gender analysis and women's experience of organizing for sustainable development. *Gender and Development, 10*(2), 21–29.

Davis, S. J., Peters, G. P., & Caldeira, K. (2011). The supply chain of CO_2 emissions. *Proceedings of the National Academy of Sciences, 108*(45), 18554–18559.

Dickinson, R. E. (1996). Climate engineering a review of aerosol approaches to changing the global energy balance. *Climatic Change, 33*(3), 279–290.

Dodman D. Blaming cities for climate change? (2009). An analysis of urban greenhouse gas emissions inventories. *Environment and Urbanization, 21*(1), 185–201.

EPA. (2016). *Global greenhouse gas emissions data*. Environmental Protection Agency. https://www.epa.gov/ghgemissions/global-greenhouse-gas-emissions-data#Sector. Accessed 20 Nov 2016.

Ferraro, A. J., Highwood, E. J., & Charlton-Perez, A. J. (2014). Weakened tropical circulation and reduced precipitation in response to geoengineering. *Environmental Research Letters, 9*(1), 014001. http://iopscience.iop.org/article/10.1088/1748-9326/9/1/014001/meta. Accessed 22 Nov 2016

Fink, J. H. (2013). Geoengineering cities to stabilise climate. *Proceedings of the Institution of Civil Engineers: Engineering Sustainability, ES5*, 242–248.

Forster, P., et al. (2007). Changes in atmospheric constituents and in radiative forcing. In S. Solomon et al. (Eds.), *Climate Change 2007: The physical science basis. Contribution of Working Group I to the Fourth Assessment Report of the Intergovernmental Panel on Climate Change*. Cambridge: Cambridge University Press.

Gardiner, S. M. (2004). Ethics and global climate change. *Ethics, 114*(3), 555–600.

Ge, M., et al. (2014). *6 graphs explain the world's top 10 emitters*. World Resource Institute. http://www.wri.org/blog/2014/11/6-graphs-explain-world's-top-10-emitters. Accessed 22 Nov 2016.

Gerber, P. J., Henderson, B., & Makkar, H. P. (2013). *Mitigation of greenhouse gas emissions in livestock production. A review of technical options for non-CO_2 emissions*. Rome: FAO.

Grasswick, H. (2014). Understanding epistemic normativity in feminist epistemology. In J. Matheson & R. Vitz (Eds.), *The ethics of belief: Individual and social* (pp. 216–243). Oxford: Oxford University Press.

Habtezion, S. (2013). *Overview of linkages between gender and climate change*. United Nations Development Programme. http://www.undp.org/content/dam/undp/library/gender/Gender%20and%20Environment/PB1-AP-Overview-Gender-and-climate-change.pdf. Accessed 24 Nov 2016.

Heede, R. (2014). Tracing anthropogenic carbon dioxide and methane emissions to fossil fuel and cement producers, 1854–2010. *Climatic Change, 122,* 229. https://doi.org/10.1007/s10584-013-0986-y. Accessed 12 Dec 2016.

Heinonen, J., & Junnila, S. (2011). A carbon consumption comparison of rural and urban lifestyles. *Sustainability, 3,* 1234–1249.

Hessing, M. (1993). Women and sustainability: Ecofeminist perspectives. *Alternatives Journal, 1*(19), 14.

Hodder, I. (1997). *Interpreting archaeology: Finding meaning in the past.* New York: Psychology Press.

Irvine, P. J., Ridgwell, A., & Lunt, D. J. (2010). Assessing the regional disparities in geo-engineering impacts. *Geophysical Research Letters, 37*(18), L18702. https://doi.org/10.1029/2010GL044447 Accessed 21 Jan 2017.

Izrael, Y. A., et al. (2009). Field experiment on studying solar radiation passing through aerosol layers. *Russian Meteorology and Hydrology, 34*(5), 265–273.

Jones, A., et al. (2010). Geoengineering by stratospheric SO_2 injection: Results from the Met Office HadGEM2 climate model and comparison with the Goddard Institute for Space Studies Model. *Atmospheric Chemistry and Physics, 10,* 5999–6006.

Keith, D. W., Parson, E., & Morgan, M. G. (2010). Research on global sun block needed now. *Nature, 463*(7280), 426–427.

Kellert, S. H., et al. (2006). *Scientific pluralism (Minnesota studies in the philosophy of science).* Minnesota: Minnesota Press.

Key, N., & Tallard, G. (2012). Mitigating methane emissions from livestock: A global analysis of sectoral policies. *Climatic Change, 112*(2), 387–414.

Khondker, H. H., & Schuerkens, U. (2014). *Social transformation, development and globalization.* Sociopedia.isa, International Sociological Association. https://doi.org/10.1177/205684601423.

Kourany, J. A. (2003). A philosophy of science for the twenty-first century. *Philosophy of science, 70*(1), 1–14.

Kovács, Á. (2012). Gender in the substance of chemistry, Part 1: The ideal gas. *HYLE – International Journal for Philosophy of Chemistry, 18*(2), 95–120. http://hyle.org/journal/issues/18-2/kovacs1.htm. Accessed 19 Dec 2016.

Kravitz, B., et al. (2013). Climate model response from the geoengineering model intercomparison project (GeoMIP). *Journal of Geophysical Research-Atmospheres, 118,* 8320–8332.

Lascano, C. E., & Cardenas, E. (2010). Alternatives for methane emission mitigation in livestock systems. *Revista Brasileira de Zootecnia, 39,* 175–182.

Leichenko, R. (2011). Climate change and urban resilience. *Current opinion in environmental sustainability, 3*(3), 164–168.

Longino, H. E. (1992). Taking gender seriously in philosophy of science. *PSA, 2,* 333–340.

Longino, H. (1994). In search of feminist epistemology. *Monist, 77,* 472–485.

Longino, H. (1995). Gender, politics and the theoretical virtues. *Synthese, 104*(3), 383–397.

Longino, H. E. (1996). Cognitive and non-cognitive values in science: Rethinking the dichotomy. In L. H. Nelson & J. Nelson (Eds.), *Feminism, science, and the philosophy of science* (pp. 39–58). Dordrecht: Kluwer Academic.

Longino, H. E. (2008). Values, heuristics, and the politics of knowledge. In M. Carrier, D. Howard, & J. Kourany (Eds.), *The challenge of the social and the pressure of practice, science and values revisited* (pp. 68–86). Pittsburg: University of Pittsburg Press.

Longino, H. E. (2013a). *Studying human behavior: How scientists investigate aggression and sexuality.* Chicago: University of Chicago Press.

Longino, H. (2013b). Subjects, power and knowledge: Description and prescription in feminist philosophies of science. In L. Alcoff & E. Potter (Eds.), *Feminist epistemologies* (pp. 101–120). New York: Routledge.

Longino, H. E., & Lennon, K. (1997). Feminist epistemology as a local epistemology. Proceedings of the Aristotelian Society, Supplementary Volumes, 71, 19–54.

Longino, H. (2010). Feminist epistemology at Hypatia's 25th anniversary 1. *Hypatia, 25*(4), 733–741.

MacGregor, S. (2009). A stranger silence still: The need for feminist social research on climate change. *The Sociological Review, 57*(s2), 124–140.

McMichael, P. (2009). Contemporary contradictions of the global development project: Geopolitics, global ecology and the 'development climate'. *Third World Quarterly, 30*(1), 247–262.

Morrow, D. R., Kopp, R. E., & Oppenheimer, M. (2009). Toward ethical norms and institutions for climate engineering research. *Environmental Research Letters, 4*(4), 045106.

Müller, B., et al. (2009). Differentiating (historic) responsibilities for climate change. *Climate Policy, 9*(6), 593–611.

Nath, P. K., & Behera, B. (2011). A critical review of impact of and adaptation to climate change in developed and developing economies. *Environment, development and sustainability, 13*(1), 141–162.

NCAR/UCAR. (2017). *New approach to geoengineering simulations is significant step forward.* NCAR/UCAR AtmosNews. Available at: https://www2.ucar.edu/atmosnews/news/129835/new-approach-geoengineering-simulations-significant-step-forward. Accessed 6 May 2018.

Nelson, V., Meadows, K., Cannon, T., Morton, J., & Martin, A. (2002). Uncertain predictions, invisible impacts, and the need to mainstream gender in climate change adaptations. *Gender & Development, 10*(2), 51–59.

Neumayer, E. (2000). In defence of historical accountability for greenhouse gas emissions. *Ecological economics, 33*(2), 185–192.

Nhanenge, J. (2011). *Ecofeminism: Towards integrating the concerns of women, poor people, and nature into development.* Lanham: University Press of America.

Niemeier, U., Schmidt, H., Alterskjær, K., & Kristjánsson, J. E. (2013). Solar irradiance reduction via climate engineering: Impact of different techniques on the energy balance and the hydrological cycle. *Journal of Geophysical Research: Atmospheres, 118*(21), 11905–11917.

NSSO. (2011). *National accounts statistics.* Government of India, Ministry of Statistics and Programme Implementation. http://www.mospi.gov.in/publication/national-accounts-statistics-2011. Accessed 1 Dec 2016.

Rasch, P. J, Tilmes, S, Turco, R. P, Robock, A Oman, L, Chen, C, Stenchikov, G. L, & Garcia, R. R. (2008, November). An overview of geoengineering of climate using stratospheric sulphate aerosols. *Philosophical Transactions. Series A Mathematical, Physical, and Engineering Sciences, 366*(1882), 4007–4037.

Resurrección, B. P. (2013). Persistent women and environment linkages in climate change and sustainable development agendas. *Women's Studies International Forum, 40*, 33–43 Pergamon.

Ricke, K. L., et al. (2012). Effectiveness of stratospheric solar-radiation management as a function of climate sensitivity. *Nature Climate Change, 2*, 92–96.

Roberts, J. T., & Parks, B. (2006). *A climate of injustice: Global inequality, North-South politics, and climate policy.* Cambridge, MA: MIT Press.

Robinson, J. (2011). Cities in a world of cities: The comparative gesture. *International Journal of Urban and Regional Research, 35*(1), 1–23.

Robock, A., Oman, L., & Stenchikov, G. L. (2008). Regional climate responses to geoengineering with tropical and Arctic SO_2 injections. *Journal of Geophysical Research: Atmospheres, 113*, D1601–D16101. https://doi.org/10.1029/2008JD010050.

Roy, A. (2009). The 21st-century metropolis: New geographies of theory. *Regional Studies, 43*(6), 819–830.

Satterthwaite, D. (2008). Cities' contribution to global warming: Notes on the allocation of greenhouse gas emissions. *Environment & Urbanization, 20*(2), 539–549.

Saurin, J. (1996). International relations, social ecology and the globalisation of environmental change. In J. Volger & M. F. Imber (Eds.), *The environment and international relations, global environmental change series.* London: Routledge.

Scheffran, J., et al. (2016). *Climate engineering: Potential pathways, consequences and risks.* Hamburg: Research Group Climate Change and Security (CLISEC), KlimaCampus, University of Hamburg.

Schwartz, R. D. (2005). Global dimming: Clear-sky atmospheric transmission from astronomical extinction measurements. *Journal of Geophysical Research: Atmospheres, 110*, D14.

Sedjo, R. A. (2010). *Adaptation of forests to climate change.* Resources for the Future, DP, 10-06.

Sheppard, E., et al. (2013). Urban pulse – Provincializing global urban- ism: A manifesto. *Urban Geography, 34*(7), 893–900.

Spanier, B. (1995). *Im/partial science: Gender ideology in molecular biology.* Bloomington: Indiana University Press.

Steinfeld, H., et al. (2006). *Livestock's long shadow.* FAO: Rome.

Strickland, S. (2012). Feminism, postmodernism and difference. In K. Lennon & M. Whitford (Eds.), *Knowing the difference: Feminist perspectives in epistemology* (pp. 265–274). New York: Routledge.

Trenberth, K. E., & Dai, A. (2007). Effects of Mount Pinatubo volcanic eruption on the hydrological cycle as an analog of geoengineering. *Geophysical Research Letters, 34*, 15. http://onlinelibrary.wiley.com/doi/10.1029/2007GL030524/full. Accessed 1 Nov 2016.

Victor, D. G, et al. (2009, March–April). *The geoengineering option: A last resort against global warming?* Geoengineering. Council on Foreign Affairs. Accessed 19 Nov 2016.

Wallerstein, I. (1992). The concept of national development, 1917–1989: Elegy and requiem. *American Behavioral Scientist, 35*, 517–529.

Warrick, R. A., & Ahmad, Q. K. (2012). *The implications of climate and sea-level change for Bangladesh.* New York: Springer Science & Business Media.

Wyckoff, A. W., & Roop, J. M. (1994). The embodiment of carbon in imports of manufactured products: Implications for international agreements on greenhouse gas emissions. *Energy Policy, 22*(3), 187–194.

Chapter 5
Novelty

Abstract This Chapter poses the question of whether sulphate geoengineering research meets the criteria of novelty as articulated by Helen Longino's feminist contextual empiricism. It includes identifying the main attributes of novelty in science, the kinds of novelty that do exist in geoengineering research, and, by articulating what genuinely novel research should look like and why it matters, and concludes that, in the context of climate engineering science, it is wanting.

Keywords Geoengineering · Feminist contextual empiricism · Novelty · Biodiversity · Paleoecology · Fractal

Novelty is the third of Longino's theoretical virtues which aims at developing theories, assumptions and methods that depart from those traditionally recognized. The hope is that such theories will not only articulate and reflect feminist commitments, which include those of justice, inclusiveness, and equity, but also make older, less progressive, and often androcentric, commitments more explicit. In this Chapter, I pose the question of whether sulphate geoengineering research meets the criteria of novelty as articulated by Longino's FCE. After defining its main attributes, I examine the kinds of novelty that do exist in this research, and, by articulating what genuinely novel research should look like and why it matters, conclude that in the context of climate engineering it is wanting. The importance of this exercise lies not only in providing a 'novel' critique of climate engineering from the perspective of feminist epistemology, but also in making the argument that these principles are of significant importance and "may properly influence scientific method and theory choice" (Anderson 1995a, b, 28).

Novelty can be juxtaposed to the Kuhnian value of consistency and/or conservatism. It functions both a "theoretical and explanatory principle", that aim to protect "against unconscious perpetuation of the sexism and androcentrism of traditional theorizing or of theorizing constrained by a desire for consistency with accepted explanatory models" (Lennon and Longino 1997, 21). This might involve the revaluation of marginalized perspectives, the upending of accepted theories with divergent viewpoints, or the introduction of new frameworks of understanding (Lennon and Longino 1997, 21).

T. Sikka, *Climate Technology, Gender, and Justice*, SpringerBriefs in Sociology,
https://doi.org/10.1007/978-3-030-01147-5_5

Novelty can come about in either a weak or strong form wherein "a strong commitment to novelty might require a complete rejection of traditional frameworks and theories, whereas a weak commitment might require reinterpretive work within existing frameworks and theories" (Tobin 2005, 127–128). It can also be results or theory driven. Two widely cited examples of novelty in science are insulin research and novel interpretations of the function of menstruation. In the case of insulin research, in the 1970s feminists challenged the traditional assumption that lengthy hormonal cycling made it economically unviable to include women in experimentation (Tavris 1992). However, as Sharon Clough argues, feminist scientists introduced novelty into the equation by questioning the accepted theory that hormonal cycling deviates from the norm and demonstrating that, in fact, "Insulin is one of the hormones that don't cycle." They also asked, "What would happen if we studied female animals and made the absence of hormonal cycling the aberration? What if we actively researched why it is that males don't menstruate?" (Clough 2004, 112). These questions introduce novelty into the equation by challenging the very assumption that insulin cycles in women. It also does so in a more theoretically radical or strong form by seeking to upend basic governing norms and assumptions. It is also significant, as an aside, that the two teams working on how to synthesize human insulin, one corporate (Genentech) the other academic (Harvard, University of California at San Francisco) took divergent paths wherein the latter group, whose interest was in high scientific content, "focused on other living beings (e.g. rats) to be studied before humans and chose a longer, more time-consuming path...with greater scientific content and expected novelty" (Fini and Lacetera 2010, 5). This indicates quite clearly the extent to which values both permeate and guide science. Overall, this particular example represents a results-driven and significant instance of scientific novelty.

With respect to menstruation, Margie Profet, in an article titled, 'Menstruations as a Defense Against Pathogens Transported as Sperm,' makes the compelling and controversial case that rather being a function-less product of bodily cycles, menstruation should be viewed as an evolutionary adaptation that "functions to protect the uterus and oviducts from colonization by pathogens," transported by, for instance, sperm (Profet 1993, 335). The research and analysis Profet relies on to reach this hypothesis is complicated and, without going into too much detail, essentially contends that menstruation must be a functional adaptive mechanism because of its precision and efficacy at purging pathogens. The strong novelty of this approach is centred on the severing of menstruation from pregnancy, sex, and motherhood, which are traditionally seen as inextricably linked. Potter argues that the notion that menstruation has an immunological function works to correct a central gender bias that interprets women's bodily processes solely under the rubric of reproduction. As a result, "cultural assumptions and bases about women's sexuality, motherhood, and the maternal-fetal relationship" (Potter 2006, 14) come to be seen androcentrically, wherein an immunological approach is never even considered. Profet's immunological theory of menstruation is both theoretically and methodological novel in a variety of ways including her innovative reading of medical literature to track uterine bleeding to infections (Profet 1993, 371). Clough nicely summarizes this novelty by arguing that, "Profet has described the function of menstruation in a way that newly synthesizes a variety of immunological and physiological research previously thought unrelated" (Clough 2003, 139).

Also of significance is precisely where novelty emerges from in this framework. Although Longino's epistemology is social, she is clear that novelty can and does come from individual experience – thereby rejecting the idea of the social as transcendental, but stressing that knowledge is product of social interactions. She states that "Without individuals there could be no knowledge; it is through their sensory system that the natural world enters cognition; it is their proposals that are subject to critical scrutiny by other individuals, their imaginations which generate novelty" (Longino 1994a, b, 143). This is noteworthy in that individual initiative, mediated through social scientific practice, is seen as both a marker and impetus to novelty in science.

Climate science in general, and the kinds of modeling and simulations associated with geoengineering specifically, do have an established ethos and history of novel theories being subject to critical scrutiny by the scientific community because of its roots in natural science. Yet, as noted, the rational force of scientific argumentation gains validation when subjected to norms produced in the social realm. Thus, it is from this social realm that the core of actionable novel ideas can emerge.

The question of whether geoengineering meets the criteria of novelty, whether hard or soft, methodological or theoretical, is complicated and must be assessed through several frames. Generally speaking, and on first glance, geoengineering has been seen as an outlandish solution to climate disruption more akin to science fiction. Schemes which include space-based reflectors or biological entities engineered to suck carbon out of the atmosphere have been described alternatively as audacious and extreme to catastrophic, risky, and a dereliction of responsibility to future generations (Goodell 2010; Keith 2000; Hamblin 2013; Hulme 2014).

Yet, it can be considered as an example of strong novelty, in that it eschews both conservatism, by pushing for a radical techno-scientific approach to climate change if mitigation and energy efficiency is insufficient, as well as consistency, by challenging other more traditional frameworks. Kravitz et al.'s work also represents novel research methodologically in that, as an intercomparison project, it aggregates multiple pieces of research. In past studies, they argue, "incomparability was limited, either due to models performing different experiments...or only a small number of models being included, each showing different results" (Kravitz et al. 2013, 8320). Their approach, the Coupled Model Intercomparison Project Phase 5 (CMIP5) addresses these concerns. Yet, it is important to make clear that a significant part of geoengineering research is interested in normalizing this novelty in order to garner public support and acceptance.

In particular, engagement with the media by the scientific communities that support further research into and potential use of geoengineering has become a central part of acquiring consent, while also playing up the excitement associated with novel science. A recent example of this can be found in a January 9, 2017 article published by *Science* which announces a recommendation made by scientists within the US government office overseeing climate science for further research into geoengineering in a move that, it argues, "will likely further discussion of deliberate tinkering with the atmosphere to cool the planet, and of directly collecting carbon from the sky, both topics once verboten in the climate science community" (Kintisch 2017). They call for ethical and responsible experimentation in the areas of CO_2 removal and storage as well as research into the dynamics of natural analogues like volcanic eruptions that could lower global temperatures. The scientists involved also make clear the "need to understand the possibilities, limitations, and potential side effects of climate intervention" which will becomes "all the more apparent with the recognition that other countries or the private sector may decide to conduct intervention experiments independently from the U.S. Government," while also emphasizing that "climate intervention cannot substitute for reducing greenhouse gas emissions" (U.S. Global Change Research Program 2017, 37). Accompanying this sober contextualization, however, is an associated "palpable sense of excitement [amongst scientists] about running experiments on computer models as well as on the environment itself" (Stigloe 2015).

In the context of stratospheric sulphate geoengineering, a myriad of innovative options related to delivery systems, from towers, ships, artillery shells, aircraft or even missiles, have been suggested which furthers the competitive excitement of big science which can be traced back to such promethean projects as missile defense, space exploration, the Internet, medical interventions, and transportation. Overall, as regards thinking outside the box as a basic component of strong novelty – this criterion has been met. Also satisfied is the 'innovative solution to a persistent problem' criteria as well as novelty understood as consistent with a "high risk/high gain" (Arrow 1962) approach to science. Climate engineering relies on an ethos of risk associated with innovation and the unknown and is discursively constructed as worthwhile in the long run in a manner similar to how genetic engineering, biotechnology and nanotechnology have been justified (Luokkanen et al. 2013; Keith 2013; MacCracken 2009; Caldeira and Keith 2010). Finally, sulphate geoengineering research can also be considered novel in that it has inspired interdisciplinary research in a variety of domains including ethics, social justice, governance, politics, as well as science and technology studies (Gardiner 2011; Barrett 2014; Robock 2008a, b).

On the other hand, there is a distinction to be made between novel perspectives and innovative technical thinking as opposed to novelty in relation to methodologies and modes of thought for which novelty should, as a derivative and non-cognitive virtue, also be responsive to: "the considerations that underlie the desire for more rigorous testing" (Lacey 2005, 215). These include multiple points of view (e.g. non-Western and feminist perspectives), novel values and standards (ethico-political and scientific) to assess research, and knowledges "historically excluded from conversation of science" (Hess 1997, 47) such as knowledge that makes gender oppression visible. The latter precept, as Hess notes, is similar to the idea of situated knowledge discussed by Sandra Harding.

Overall, and as a practical endeavour, introducing novelty into climate engineering science on a methodological level is considerably more complicated. Yet, some direction can be taken from recent climate change research which has succeeded in incorporating innovative standpoints and methodologies. Creative methodologies, like new forms of modeling and simulation, as well as novel perspectives like paleo-ecology and solar science, have been in gaining recognition. A convincing example can be seen in projects to generate more models and climate datasets on a local level that retain the high resolution that are a feature of global models. According to research conducted in 2016 by scientists at NASA, such initiatives as topographic downscaling can "add value to intermediately downscaled maximum and minimum 2-m air temperature at high elevation stations, as well as moderately improving domain-averaged maximum and minimum 2-m air temperature," while improving mean precipitation forecasts as well (Winter et al. 2016, 881). This approach could prove useful to climate engineering research for which precipitation levels are notoriously hard to account for (remember that fluctuating precipitation, more droughts and floods, is one conceivable side effect of sulphate geoengineering).

On the level of perspective, paleoecology, which focuses on the study of the earth's geological and biological history using fossils, has been proposed as a significant area of scientific research that can contribute to climate science in novel ways. As Pardi and Smith contend, "Analysis of the paleontological record can yield valuable information on how past climate change has shaped biodiversity in the past, and provide clues for what we may expect in the future" (Pardi and Smith 2012, 93). Paleoecological research has allowed lab and field studies on animals to be integrated into climate science and could also facilitate new kinds of geoengineering research by inserting the subject of animal and plant biodiversity into the equation.

The issue of biodiversity in relation to climate engineering has been taken up by the Convention on Biological Diversity (CBD), whose governing body has been hesitant to endorse anything but small scale and controlled climate engineering research. Their mandate is to protect biodiversity's current and future state – which is threatened by climate change, and, by their own admission, potentially by climate engineering. (Sugiyama and Taishi 2010; Kwa and Hemert 2011). The CBD's position on sulphate climate engineering brings the neglected concern of biodiversity into the light by noting the potential for unintended consequences including, for instance, the possibility of ozone depletion and ocean acidity (Williamson et al. 2012). Genuine novelty would require that geoengineering models actually incorporate biodiversity loss into their models or create ones that 'center' biodiversity specifically. Scientists working on 'climate envelope modelling' have introduced this approach. Climate envelope modelling has been used to project probabilistic representations of current species distribution onto present day climate, as a baseline, and subsequently at different temperature scenarios (Warren et al. 2013). Nothing of a similar nature has been done with respect of geoengineering. Sulphate geoengineering in particular could benefit immensely from models that thematize biodiversity.

Methodologically, gender effects, as in the biodiversity example, could also be built into climate and climate engineering models in ways that go beyond simply presenting the model's findings and adding contextualizing information. This traditionally consists of, for example, commenting on how decreased climate disruptions might help women as a generalized category (e.g. by making dislocation less likely, resulting in more stable access to food and water, and fewer economic disruptions). Building gender into models has been successfully used in the discipline of geography to construct geographical information systems (GIS) that merge experiential knowledge with mapped representations to, for instance, map the observation of increased rates of breast cancer in a particular region in a way that allows for the acquisition of "knowledge outside the realm of daily experience and for connecting their [the community's] personal experiences of health and illness to a wider social and political agenda" (McLafferty 2002, 266). It is this kind of grassroots led research that builds local concerns into models which is virtually nonexistent in the area of climate geoengineering.

On the level of epistemology, there are concerns related to the top down nature of geoengineering research in that it focuses primarily on what *we* may or may not do to tinker with the climate, how that climate would react to *our* actions, and how these reactions might affect *our* quality of life. In doing so, this framework employs a prototypically humanist, capitalist, and hierarchical understanding of the world consistent traditional, Western science. Keeping with epistemology, a feminist approach to this limiting bias would introduce novel factors and issues including approaching the world's climate more intimately by "shorten[ing] the distance between the observer and the object being studied" and by considering "the complex relationship between the organism [us] and the environment" (Tuana 1989, 9). Concerns arising from a critical feminist perspective would propose novel questions like,

> How did the existing relationship originate? Which classes and groups are benefited by the existing relationship? How does the dominant coalition maintain and perpetuate its control? What are the consequences of present distribution of power in the organization for present society and for future generations? (Marshall 1987, 18).

While treatments of geoengineering by social science, policy studies, and political science have attended to some of these subjects, science driven studies have not – and particularly not as a feminist act. Inserting conceptualizations of science driven by feminist concerns might encourage studies that disrupt traditional methods in both climate and geoengineering science, introduce more variability, play with parameterization, and encourage local scales and variables. This is particularly the case since, as Hilary Rose maintains, fundamental physics, on which all models of climate engineering rely, is "at once the most arcane and the most deeply implicated in the capitalist system of domination" (Rose 1994, 8). Its assumptions, therefore, can and should be tested and critiqued.

Fractal climate response functions and nonlinear methodologies constitute two established, yet still novel, starting points for more reflexive and innovative climate engineering research. The fractal approach deals with multiple time series and data streams rather than large blocks of time (e.g. seasons versus months). In doing so, researchers are able to integrate temperature and precipitation in order "to define an two-dimensional data stream" (Nunes et al. 2011, 53) that aids in revealing patterns and making connections between climate variables (Hateren 2013).

This approach builds on fractal frameworks used in recent sociological and feminist work that incorporates chaos, dynamism, and patterns into science in manner meant to reveal what has been previously hidden and, in doing so, raise questions that include why and to what effect. The use of a fractal theory in the example given above makes inroads into incorporating novelty, yet does not take advantage of what feminist approaches to fractalization has made possible on a methodological level. Cheryl Hercus, for instance, uses fractals to draw out the iterative and recursive structure of subjectivity that she contends is structured by discourses and practices that incorporate "ways of knowing, feeling, belonging, and doing" (Hercus 2005, 10). This is a departure from a Cartesian conception of subjectivity. Stratospheric SRM geoengineering research, to be truly novel, could incorporate fractal modelling in order to integrate variability and conceptual difference reflected in competing notions of time, space, perspective, power, inequality, race, gender and class.

Furthermore, sulphate climate engineering research would benefit from a flexible research methodology that incorporates previously discarded details and potential side effects – and discusses them in more detail; thematizes the differences and the interplay between the local and the global, the micro and the macro; engages with the dynamics of scale and perspective; and pays attention to relationships between people, institutions, and objects (Strathern 1989; Kellert 1996).

A further novel approach that binds epistemology and methodology in manner consistent the feminism deployed by FCE, would see sulphate geoengineering examined not solely in relation to parameterized idealized systems, but also in terms of interaction and bodily experience or embodiment. Embodied knowledge can be defined as felt knowledge derived from bodily experience and intuition and has been used, in the scientific context, to study knowledge of such phenomenon as disability and pregnancy, (Cheyney 2008; Scully and Mackenzie 2007). If we think of embodiment as central to ethical subjectivity, it should also be considered as a significant aspect of epistemology in the so-called 'hard sciences' as well.

Fundamentally, novel research into climate engineering requires critical questioning and openness. An embodied approach might involve on the ground fieldwork to garner knowledge of how people are experiencing climate change and how they might feel about interventions that come with a whole host of potential side effects. These communities will also have significant stores of knowledge grounded in experience that "go beyond disembodied information, cool cognition, and cultural schemas that appear independent of individual experience" (DiMaggio 1997, 273). Engaging in research methodologies that focus on the knowledge and practices of those most potentially affected by climate interventions (indigenous communities, racialized groups, women) allows for new questions to be asked based on novel information gathered from tacit practices that involve real life engagement with the environment on a day-to-day basis (Polanyi 1966).

This requires engaging in fieldwork rooted in ethnographic approaches that might involve dynamic participation or observation. Concerns around how communities might react and feel about sulphate geoengineering induced disrupted monsoons, ocean acidification and ozone depletion, which could also disturb fishing and food supplies, and what these communities (primarily in the Global South but also in agricultural and coastal communities in more affluent regions) think about the difficulties associated with terminating climate interventions, are significant. As Colin Peile argues, "Knowledge talked about in this way leads to an appreciation of non-conceptual forms of knowledge held in feelings, actions and behaviours rather than the mind" (Peile 1998, 49). This knowledge could then filter into basic research and, potentially, transform research priorities, objectives, and conclusions.

Finally, it is also the case that because new forms of embodied knowledge are experienced through material and social contexts, aesthetic concerns associated with embodiment must also be considered. In regards to sulphate geoengineering, the process that requires the injection of particulates into the atmosphere would likely alter our skies and horizons making blue skies more white. The cultural, social, and psychological significance of blues skies should not be underestimated (Robock 2008a, b). Anna North, in an insightful 2015 op-ed for the *New York Times*, draws on Dacher Keltner's research on the positive health effect of awe by making a suggestive correlation between experiences in the natural world and better physical and emotional health. Keltner argues that without a sky in which we can see the stars and a blue horizon, the cognitive and emotional reflectivity associated with experiencing nature is lost: "kids are going to be less imaginative, we're going to be less modest and less kind to each other," and "it may cost us in terms of health" (North 2015). This perspective constitutes the kind of novelty a feminist approach like FSE might center in scientific research.

Cumulatively, as evidenced by the range of limitations presented in this Chapter, it is clear that, while climate geoengineering in general, and sulphate climate engineering in particular, represents a novel suite of solutions to climate disruption in terms of formal originality and inventiveness consistent with a technophilic culture, it does not meet the criteria of feminist novelty as articulated by Longino. I have discussed at length what this kind of research might look like – including everything from a more open methodology to a fractal approach, a focus on embodiment rooted in ethnographic field work, and the consideration of aesthetics. Novelty, intrinsically, is a virtue to be strived for since its realization, when coupled with the other virtues, works to reveal inequalities and power asymmetries of various kinds. It also "privilege[s] theories that postulate different entities, adopt different principles of explanation, or investigate what traditional scientific inquiry has not" (Barker 2004, 222). When interpreted in this way, it becomes clear that, stratospheric sulphate geoengineering research does not quality as novel research. Longino's third feminist virtue, mutuality of interaction, introduces a further element of research lacking in contemporary climate engineering research that I explore in the next Chapter.

References

Anderson, E. (1995a). Feminist epistemology: An interpretation and a defense. *Hypatia, 10*(3), 50–84.
Anderson, E. (1995b). Knowledge, human interests, and objectivity in feminist epistemology. *Philosophical Topics, 23*(2), 27–58.

Arrow, K. (1962). Economic welfare and the allocation of resources for invention. In R. R. Nelson (Ed.), *The rate and direction of inventive activity: Economic and social factors* (pp. 609–629). Princeton University Press: Princeton.

Barker, D. K. (2004). From feminist empiricism to feminist poststructuralism: Philosophical questions in feminist economics. In J. B. Davis & A. Marciano (Eds.), *The Elgar companion to economics and philosophy* (pp. 213–230). Cheltenham: Edward Elgar Publishing.

Barrett, S. (2014). Solar geoengineering's brave new world: Thoughts on the governance of an unprecedented technology. *Review of Environmental Economics and Policy, 8*(2), 249–269.

Caldeira, K., & Keith, D. W. (2010). The need for climate engineering research. *Issues in Science and Technology, 27*(1), 57–62.

Cheyney, M. J. (2008). Homebirth as systems-challenging praxis: Knowledge, power, and intimacy in the birthplace. *Qualitative Health Research, 18*(2), 254–267.

Clough, S. (2003). *Beyond epistemology: A pragmatist approach to feminist science studies.* Lanham: Rowman & Littlefield Publishers.

Clough, S. (2004). Having it all: Naturalized normativity in feminist science studies. *Hypatia, 19*(1), 102–118.

DiMaggio, P. (1997). Culture and cognition. *Annual Review of Sociology, 1*, 263–287.

Fini, R., & Lacetera, N. (2010). Different yokes for different folds: Individual preferences, institutional logics, and the commercialization of academic research. In G. D. Libecap et al. (Eds.), *Spanning boundaries and disciplines: University technology commercialization in the idea age* (pp. 1–26). Bingley: Emerald Group Publishing.

Gardiner, S. M. (2011, May 1). Some early ethics of geoengineering the climate: A commentary on the values of the Royal Society report. *Environmental Values, 20*(2), 163–88.

Goodell, J. (2010). *How to cool the planet: Geoengineering and the audacious quest to fix earth's climate.* Chicago: Houghton Mifflin Harcourt.

Hamblin, J. D. (2013). *Arming mother nature: The birth of catastrophic environmentalism.* London: Oxford University Press.

Hercus, C. (2005). *Stepping out of line: Becoming and being feminist.* Oxford: Psychology Press.

Hess, D. J. (1997). *Science studies: An advanced introduction.* New York: New York University Press.

Hulme, M. (2014). *Can science fix climate change: A case against climate engineering.* New York: Wiley.

Keith, D. W. (2000). Geoengineering the climate: History and prospect. *Annual Review of Energy and the Environment, 25*(1), 245–284.

Keith, D. (2013). *A case for climate engineering.* Cambridge, MA: MIT Press.

Kellert, S. H. (1996). Science and literature and philosophy: The case of chaos theory and deconstruction. *Configurations, 4*, 215–232.

Kintisch, E. (2017). U.S. should pursue controversial geoengineering research, federal scientists say for first time. *Science*, 9 January 2017. http://www.sciencemag.org/news/2017/01/us-should-pursue-controversial-geoengineering-research-federal-scientists-say-first. Accessed 15 Jan 2017.

Kravitz, B., et al. (2013). Climate model response from the geoengineering model intercomparison project (GeoMIP). *Journal of Geophysical Research-Atmospheres, 118*, 8320–8332.

Kwa, C., & van Hemert, M. (2011). Engineering the Planet: The issue of biodiversity in the framework of climate manipulation and climate governance. *Quaderni, 3*, 79–89.

Lacey, H. (2005). *Is science value free?: Values and scientific understanding.* Oxford: Psychology Press.

Longino, H. (1994a). In search of feminist epistemology. *Monist, 77*, 472–485.

Longino, H. (1994b). The fate of knowledge in social theories of science. In F. Schmitt (Ed.), *Socializing epistemology: The social dimensions of knowledge* (pp. 135–157). Lanham: Rowman and Littlefield Publishers Inc.

Longino, H. E., & Lennon, K. (1997). Feminist epistemology as a local epistemology. *Proceedings of the Aristotelian Society*, Supplementary Volumes, *71*, 19–54.

Luokkanen, M., Huttunen, S., & Hilden, M. (2013). Geoengineering, news media and metaphors: Framing the controversial. *Public Understanding of Science*, 1–16. http://pus.sagepub.com/content/early/2013/02/14/0963662513475966. Accessed 4 Oct 2016.

MacCracken, M. (2009). Beyond mitigation: Potential options for counter-balancing the climatic and environmental consequences of the rising concentrations of greenhouse gases. *World Bank Policy Research Working Paper Series*.

Marshall, W. M. (1987). *Environments and organizations*. London: Jossey-Bass.

McLafferty, S. L. (2002). Mapping women's worlds: Knowledge, power and the bounds of GIS. *Gender, Place and Culture: A Journal of Feminist Geography, 9*(3), 263–269.

North, A. (2015). What if we lost the sky? Opinion Pages, *New York Times*. 20 February 2015. https://op-talk.blogs.nytimes.com/2015/02/20/what-if-we-lost-the-sky/?_r=0. Accessed 19 Jan 2015.

Nunes, et al. (2011). Fractal-based analysis to identify trend changes in multiple climate time series. *Journal of Information and Data Management, Belo Horizonte, 2*(1), 51–57.

Pardi, M. I., & Smith, F. A. (2012). Paleoecology in an era of climate change: How the past can provide insights into the future. In J. Louys (Ed.), *Paleontology in ecology and conservation* (pp. 93–116). Berlin: Springer.

Peile, C. (1998). Emotional and embodied knowledge: Implications for critical practice. *Journal of Sociology & Social Welfare, 25*(4), 39–60.

Polanyi, M. (1966). *The tacit dimension*. New York: Anchor Press.

Potter, E. (2006). On the very idea of a feminist epistemology for science. *Metascience, 15*(1), 1–37.

Profet, M. (1993). Menstruation as a defense against pathogens transported by sperm. *Quarterly Review of Biology, 1*, 335–386.

Robock, A. (2008a). 20 reasons why geoengineering may be a bad idea. *Bulletin of the Atomic Scientists, 64*(2), 14–18.

Robock, A. (2008b). Whither geoengineering? *SCIENCE-NEW YORK THEN WASHINGTON, 320*(5880), 1166.

Rose, H. (1994). *Love, power, and knowledge: Towards a feminist transformation of the sciences*. Bloomington: Indiana University Press.

Scully, J. L., & Mackenzie, C. (2007). Moral imagination, disability, and embodiment. *Journal of Applied Philosophy, 24*(4), 335–351.

Stigloe, J. (2015). Can volcanoes tackle climate change? *The Guardian*, 10 April 2015. Accessed https://www.theguardian.com/environment/2015/apr/10/can-volcanoes-tackle-climate-change-frankenstein-mount-tambora

Strathern, M. (1989). Comment on 'Capitalising difference.'. *Australian Feminist Studies, 9*, 25–29.

Sugiyama, M., & Taishi, S. (2010). Interpretation of CBD COP 10 decision on geoengineering. Central Research Institute of Electric Power Industry (Japan). http://criepi.denken.or.jp/en/serc/research_re/download/10013dp.pdf. Accessed 4 Jan 2017.

Tavris, C. (1992). *The mismeasure of woman*. New York: Simon and Schuster.

Tobin, T. W. (2005). Assessing moral theories. Lessons from feminist philosophy of science. In L. N. Gurley et al. (Eds.), *Feminists contest politics and philosophy: Selected papers of the 3rd interdisciplinary conference celebrating international women's day* (pp. 125–138). Bruxelles: Peter Lang.

Tuana, N. (Ed.). (1989). *Feminism and science*. Bloomington: Indiana University Press.

U.S. Global Change Research Program. (2017). *National global change research plan 2012–2021: A triennial update*. Washington, DC.

Van Hateren, J. H. (2013). A fractal climate response function can simulate global average temperature trends of the modern era and the past millennium. *Climate Dynamics, 40*(11-12), 2651–2670.

Warren, R., et al. (2013). Quantifying the benefit of early climate change mitigation in avoiding biodiversity loss. *Nature Climate Change, 3*(7), 678–682.

Williamson, P., et al. (2012). Impacts of climate-related geoengineering on biological diversity. Part I of: Geoengineering in relation to the convention on biological diversity: technical and regulatory matters, Secretariat of the Convention on Biological Diversity, Montreal, Technical Series, 66.

Winter, J. M., et al. (2016). Development and evaluation of high-resolution climate simulations over the mountainous northeastern United States. *Journal of Hydrometeorology, 17*(3), 881–896. https://doi.org/10.1175/JHM-D-15-0052.1.

Chapter 6
Mutuality of Interaction

Abstract This Chapter discusses the mutuality of interaction or complexity of relationship virtue. An overview of its conditions is provided as well as examples of where this kind of process-oriented science can be found. Of all the theoretical virtues, this is the one that the science of sulphate geoengineering has come closest to fulfilling Yet, it pointed out that there are aspects of this particular virtue not fully taken advantage of as a result of the way geoengineering science is structured.

Keywords Interaction · Mutuality · Feminist virtues · Feminist science · Midwifery · Networks · Agential · Feedback loops

This Chapter discusses the third of Longino's feminist scientific virtues – namely, mutuality of interaction or complexity of relationship. In addition to offering an overview of its conditions, also examined are examples of where this kind of process oriented science can be found, and steps that have been made to incorporate it into geoengineering research. Of all the theoretical virtues, this is the one which the science that underpins sulphate geoengineering has come closest to fulfilling. However, there are aspects of this particular virtue not fully taken advantage of as a result of the way the scientific practice is structured such that it is resistant to a fulsome implementation of dialogic and reflexive methodologies and principles. Overall, this Chapter contributes a further set of arguments that reinforce the objective of this book which is to provide a novel basis on which to critique climate engineering and posit a role for feminist science.

Mutuality of interaction or complexity of relationship, as Longino and Lennon contend, involves incorporating complexity, dynamic interplay, and coevolution into scientific practice as opposed to its companion, ontological heterogeneity, "which tolerates the existence of different kinds of things, complexity, mutuality, reciprocity characterize their interactions." Thus, an interactionist approach favors theories in which interactions are "represented as complex" and involve "mutual and reciprocal relationships among factors" (Longino and Lennon 1997, 22). This requires models that are horizontal or networked rather than vertical or hierarchical. It also necessitates the rejection of,

T. Sikka, *Climate Technology, Gender, and Justice*, SpringerBriefs in Sociology,
https://doi.org/10.1007/978-3-030-01147-5_6

101

...single factor models for models that incorporate dynamic interaction, models in which no factor can be described as dominant or controlling and that describe processes in which all factors influence the others (Longino 1996, 4).

The Earth's system, as an object of study, is uniquely placed for investigation using a dynamic and relational approach since, structurally; this framework emphasizes the complex linkages between the various components of the system – i.e. the biosphere, hydrosphere, atmosphere, and cryosphere. The symbiotic relationship between these spheres, as it relates to the climate, demands a scientific practice which acknowledges that the regulation of the climate occurs through a complicated set of interactions between systems and an awareness that "The interconnectedness of Earth's systems means that a significant change in any one component of the climate system can influence the equilibrium of the entire Earth system" (US GCRP 2009). As such, climate change science lends itself to the feminist virtue of mutuality of interaction articulated FCE quite nicely.

Climate science modeling, as demonstrated further on, has come some way in developing modeling practices that incorporate process and interface in innovative ways. Yet, it is important to keep in mind that Longino's virtue requires a substantive and transformational commitment to hypotheses, theories, methods, and research questions that are founded on mutual influence and dynamic interaction. While there are movements in that direction more needs to be done to fulfill the conditions of mutuality of interaction. Meeting these criteria should ultimately result in scientific principles, findings, understandings, and outcomes that are sociopolitically and ethically progressive.

A cogent example of interactionist science given by Longino is that of midwifery. While Longino argues that midwifery is first and foremost an example of diffusion of power, in that it encourages "the individual women either to made decisions about her health or to retain control over her own body" (Longino 1996, 48), Yousefi makes the case that it is also a practical example of mutuality of interaction since it "encourages reciprocal relationships" (Yousefi 1998, 81) amongst practitioners and patients in a social context that relies on "knowing, sensing and intuiting" as much as on formal knowledge (Dimen 1989, 47). As with FCE's other virtues, apart from empirical adequacy, one of its core objectives is to reveal inequities rooted in gender and other forms of hierarchy.

The midwifery example explicitly thematizes gender, as is often the case with examples found in biological science, but, as previously articulated, gender as an explicit issue is less perceptible in the so-called hard or physical sciences. As such, it is important to reiterate that what makes Longino's approach feminist involves being committed to practicing science in a progressive way by recognizing "the collusion between scientific knowledge and categories of difference" and the "historical and political contexts of...scientific production" (Subramaniam 2014, 5). Gender, as such, may not always be unambiguously front and center, but the feminist insight that science, as a social enterprise, is value-laden, is. As Spanier states,

Ideally, feminist research originates in the material and political concerns of women-centered efforts to improve the quality of life for women, children and hence, the planet.

Women's concerns about what is wrong with society, such as violence, poverty, sexual abuse, and the misuse of power over people and resources, are placed at the center of a feminist approach, in contrast to conventional scientific motivations, such as the accrual of knowledge for its own sake, the advancement of capitalism, or personal ambition. Explicitly stated feminist values may reorder funding priorities for research, question metaphors chosen and promoted for biological life, and revision who should be recruited in to the profession (Spanier 1995, 41).

Another example that typifies mutuality of interaction or complexity of relationship, specifically in relation to the kind socio-political committedness expressed above, is conveyed by Sylvia Nagl who makes use this virtue in her study of protein domains which she treats as complex adaptive systems using the framework of neural networks. This perspective, which highlights the importance of networks and systems in biology, is particularly relevant to climate science and climate engineering. Essentially, whereas proteins have traditionally been treated as linear chains of amino acids entities, Nagl argues that the networked approach holds that each acid sequence should be thought of as agents that can assume different states. When applied to geo-engineering, a networked approach would hold that the variables involved in modeling are active, interactive, and variable. An example of research that does this can be found in Karsten Steinhaeuser et al.'s article titled 'Complex Networks as a Unified Framework for Descriptive Analysis and Predictive Modeling in Climate Science,' in which the authors use a networked approach to integrate descriptive analysis and predictive modeling in order to demonstrate how networked clusters of data can yield robust results. Underlying this perspective is the view that a,

...unified framework prompts 'networked' thinking: imagine the globe as a spatiotemporal grid. Each cell, corresponding to a region, can be represented by a node, and different nodes are connected to each other not by spatial proximity, but rather on the basis of similarity shared in climatic variability. Such interactions among nodes can be exploited to discover how regions are related and impact each other (Steinhaeuser et al. 2011, 1).

Vis-à-vis climate engineering, this kind of interactionism is present when, for example, theoretical projections and parameters are structured through the lens of networks. For example, work has been done to examine the sulphur cycle by including the interaction of natural and pre-existing sulphate aerosols and by manipulating parameters around cloud droplets and rain (Jones et al. 2001). In fact, the very act of coupling models could also be considered moves towards greater interactionism. Intercomparison projects, like the GeoIMP, are illustrative of more interplay which, according to the IPCC, has yielded,

A new era for climate modeling, setting standards of quality control, providing organisational continuity and ensuring that results are generally reproducible...[as well as] quantified improved agreement between simulated and observed atmospheric properties as new versions of models are developed (IPCC 2007).

The integration of feedback loops is another way in which complexity of relationship is evident in geoengineering research. The integration of knowledge that there will be a difference between any reduction in temperature as a result of SRM geoengineering and Polar and Equatorial regions because of the "spatial difference between the radiative forcing effect of the reduced insolation and the raised CO_2

levels...[which] is amplified by the operation of temperature feedbacks involving snow cover and sea-ice extent" (Irvine et al. 2010), is one such example. When insights like this are built into the system, it is more likely to yield fruitful results and divergent perspectives. Yet, it is also important to note that when estimating feedbacks in climate engineering, it is often the case that tests are conducted by returning to thermodynamic models that operate from an equilibrium rather than an interactionist response (Andrews et al. 2012; Williams et al. 2007). Still, most scientists recognize that this kind of complexity should be welcomed rather than avoided.

An further example of climate engineering science that incorporates reflexivity can be found in the not insignificant amount of SRM geoengineering research that concentrates on taking the generalized results of AOGCM models that run climate engineering scenarios, and sharpening their focus on the interplay between those results and one particular factor or phenomenon. In a piece written by Kuebbeler et al. (2012), the authors argued that cirrus clouds, which regulate the "radiation budget of the Earth-atmosphere system," are likely to be unintentionally thinned by stratospheric aerosols due to particular "microphysical and dynamical effects." In the course of reaching this conclusion, Kuebbeler et al. draw on the complex interplay between modeling, simulations, and historical data (from Mt. Pinatubo) to calculate that while, "Optically thinner cirrus clouds lower the planetary albedo and more short-wave (SW) radiation enters the Earth-atmosphere system, leading to a warming," it is also the case that, "optically thinner cirrus clouds trap less longwave (LW) radiation such that more terrestrial radiation is emitted to space, inducing a cooling" (Kuebbeler et al. 2012). In nod to how complicated the interplay between the climatic, atmospheric, terrestrial, and oceanic systems are, and evidenced by the lack of a definitive conclusion on their effects on cirrus clouds, the authors make clear that much more research needs to be done on, for example, the role of different sea surface temperatures and their effects on ozone chemistry. Overall, however, research like this demonstrates a predisposition towards perspectives that integrate "simultaneity, mutuality and coevolution...as defining characteristics" (Bjornerud 1997, 98).

This notion of active agential relations, involving variables that are not static but dynamic and mutable in the spirit of Nagl's work, has been drawn on by feminist science theorists like Karen Barad whose research is uniquely applicable to climate engineering. Barad uses an agential approach to examine the practice of physics which is relevant to modeling climate engineering since it too relies on basic physical laws around light, radiation balance, carbon balance, the thermodynamics of gas, the kinetic energies of molecules, the microphysics of clouds etc.

Barad draws on Niels Bohr's conception of quantum physics to challenge reductionism, promote interactionism, reassert the primacy of the relationship between us and the natural world, emphasize the active role instruments play in the course of scientific practice, and buttress her thesis that,

> ...scientific practices must therefore be understood as interactions among component parts of nature and that our ability to understand the world hinges on our taking account of the

fact that our knowledge making practices are social-material enactments that contribute to, and are a part of, the phenomena we describe (Barad 2007, 26).

She also argues that these kinds of 'agential relationships' imply a continuous intra-action between scientists and the subject/object being studied and between human and nonhuman actors. Barad is clear that understanding agential relationships requires "an analysis that enables us to theorize the social and the natural together, to read our best understandings of social and natural phenomena through one another in a way that clarifies the relationship between them" (Barad 2007, 25). Her grounding in feminist studies of science gives rise to the further interrogation of foundations, critique of dualisms, and elevation of performative analyses that cultivate an embodied approach to science in which, for example, particles are seen as constitutively entangled with the world, renormalization of perversions are rejected, and mutual entanglements of different kinds are reinforced as ethical adoptions of Otherness (Barad 2012).

Now, this approach may initially appear esoteric and difficult to apply to climate science and geoengineering, yet there has been movement to integrate perspectives into climate science that are similarly dynamic. To begin with, sociological examinations of climate change have been quick to integrate the idea of climate change as a socially constructed artefact by examining the complexity of the Earth's systems, the two-way relationship between scientists and nature (with respect to knowledge production), difficulties with models, the politics of representation (including the need for idealizations and parameterization), and the challenge posed by variability and extremes (Stehr and Von Storch 1995; Kirton et al. 2013, Cass and Pettenger 2007). There is also geoengineering research, that seeks to highlight the mutual relationship between hard science and perceptions or beliefs about nature, the climate, and the world as coevolutionary which contributes to how climate engineering is understood – e.g. as a necessary emergency measure or as a moral hazard etc. (Heyward and Rayner 2013; Horton 2015).

Yet, in terms of the practice of modeling and simulation, very little work has been done to integrate an understanding of the social and natural world as entangled and in which "human corporality...is inseparable from 'nature' or 'environment'" (Alaimo 2008, 238). As Neimanis and Walker argue, for an approach to do this it would require a reconceptualization of weathering in which humans, their artefacts, and weather overlap spatially and corporeally, where rain is seen to "extend into our arthritic joints...[the] sun might literally color our skin, and the chill of the wind might echo through the hidden hallways of our eardrums" (Neimanis and Walker 2014, 560). The question of how sulphate geoengineering "might change me" (Neimanis and Walker 2014, 561), and the world in sensed ways, i.e. in terms of how it might feel to live in a world with white skies and to swim in acidifying seas, should be asked of all climate intervention schemes.

In addition to Barad, there is also a significant amount of relevant work in the philosophy of mathematics and physical sciences that pose divergent understandings of, "number, size, quantity, possibility, shape, algorithmic problem solving, analogic representation, and other extended components of mathematical thinking and living" (Appelbaum 2016, 5), which could be applicable to the study of geoengineering with respect to modelling as well as basic epistemology.

This includes postcolonial epistemologies, that highlight partial and relational knowledge, aesthetics and beauty in representation (Gutiérrez 2012; Sinclair 2009), and feminist contributions to physics that seek to challenge hierarchy, and norms of objectivity (Keller 1987; Whitten 1996; Bleier 1986). Feminist approaches to physics challenge the discipline in ways that build on Longino by, for example, "reject[ing] any fundamental distinctions as system/observer. .. physical system/consciousness" (Rovelli 1996, 1647; Bug 2003), questioning what sits inside and outside the boundaries of disciplines, and harnesses contextual, reflexive, and non-objectivizing methodologies by "re-admit[ing] the human subject into the production of scientific knowledge" (Heckman 1990, 130). Postconstructivist philosophy might also contribute to geoengineering science through its efforts to "challenge one-sided accounts of scientific knowledge and foster more self-reflective research practices," by highlighting the significance of non-knowledge, scientific practice and performativity (Wehling 2006, 81).

A final notable contribution on this particular matter is Nancy Tuana's conception of interactionism and theory of viscous porosity, which complements Longino's virtue of mutuality of interaction with additional insights pertinent to geoengineering. First, Tuana characterizes interaction as removing "any hard and fast divide between nature and culture" wherein "The world is [seen to be] neither fabricated in the sense of created out of human cultural practices, nor is its existence [seen as] independent of a multitude of forms" (Tuana, 191; Tuana 2001, 238–239). Tuana draws on this basic assumption to articulate a theory of 'viscous porosity' which "refers to the resistance of matter to changing form, which makes it thus somewhat reliable as an object of knowledge, while porosity acknowledges the mutual influence of the material and the social" (Tillman 2015).

This neologism provides the frame from which to ascertain:

1. The constructed ways in which human intervention into the natural environment characterizes both nature and humanity as separate – and where one (human) acts on the other (the environment);
2. How the various equations and constructions used in AOGCMs build on the GCM's focus on physical processes, e.g. the land, the ocean, atmosphere and cryosphere, and supplements them with equations and parameterizations that integrate surface pressure and oceanic fluxes are conceived of as material and not social processes.
3. How and why the discounting of social, bodily phenomenon has become the norm and,
4. What research that perceives matter as viscous, in exerting material pressures on scientific practice, and also constituted with and by the social would look like in the context of geoengineering?

Without answers to these questions, it is difficult to characterize scientific research into solar climate engineering as interactionist in a manner consistent with the FCE approach.

At this point, it is beneficial to spend some time discussing GCMs and AOGCMs in more detail – particularly since current models do reflect a significant improvement in accounting for dynamic climatic processes. GCMs aim to simulate how the various components of the earth's system interact with each other. This includes "anthropogenic effects and coupled oceanic atmospheric, and some land-surface processes" that work by "mathematically representing the physical movement of gaseous or liquid masses, energy transfers, reflection, absorption, and other phenomena" (Lahsen 2002, 6). More advanced 3D models use a global grid and some incorporate time scales. At their core, climate models aim to represent the complex interaction between the atmosphere and the biosphere or biochemical cycles and the climate. A very basic example that exemplifies improvements in modeling is the use of discretized quantum field theory (QCD) in computer modeling which uses interactive and nonlinear calculations to move from numerical simulation to direct statistical simulation. These models are better at capturing the effects, formation, and characteristics of notoriously complex and fast moving jet streams (Tobias and Marston 2013). Examples like this push the practices underlying geoengineering science towards a more interactionist approach.

One of the most interesting techniques being used to integrate dynamism into sulphate geoengineering research has been implemented by Kravitz et al.'s GeoMIP team through the climate emulators, rather than just GCMs, to "project climate effects from different possible future pathways of anthropogenic forcing" using a variety of greenhouse gas concentrations and different amounts of sulphate geoengineering (Kravitz et al. 2016, 15,789). Most notably, the 'pattern scaling' emulator draws on a "predictive dynamic model used only for the time evolution of the global-mean temperature" which is then supplemented by "land-sea temperature contrast…functions of temperature….or more spatial degrees of freedom to better predict regional effects" (Kravitz et al. 2016, 15,790). The latter of which, as noted, is often neglected.

Understood in this way, the science used to reinforce climate and geoengineering research can seen to employ elements of an interactionist ethos, yet, as also made clear, it does not rest on a foundation of dynamic interactionism whose objective is to shape science in a socially progressive and ethically sound way. It is important, in making this argument, to delve a little deeper into what precisely makes Longino's theoretical virtue of mutuality such a critical virtue. For this I turn to Nagl who articulates specific principles of adaptive and interactionist systems that not only complements Longino's framework, but supplements it with more detail. It can therefore be used as a guidepost to shape geoengineering research towards more dynamic and interactionist practices. Nagl's principles are as follows:

1. Complex systems consist of a large number of elements;
2. The elements of a complex system interact in a dynamic fashion and these interactions change over time;
3. The interactions between elements are richly connected – any one element influences, and is influenced by, a large number of others;

4. The interactions between elements are non-linear. Small causes can have large results, and vice versa. Complexity results from the patterns of richly connected interactions between the elements;
5. The interactions between elements are relatively short-range;
6. There are recurrent interaction pathways; and
7. Complex systems have a history.

Cursorily, stratospheric sulphate geoengineering research does appear to reflect aspects the principles listed above. It does consist of research that studies a large multifaceted system; there is an orientation to treating said system as interactive, connected, and non-linear; and pathways and rich interconnections (as a result of multi-causal factors) are built into most models and simulations. However, on the subjects of history and context, there is significant room for improvement. A reflexive analysis of the context in which discussions of climate engineering is taking place, i.e. under a neoliberal economic system characterized by unequal relations of power, extreme distributive inequalities, and disruptive socio-economic vicissitudes, is needed. Also needed is a historical examination of climate engineering that integrates accounts of the genesis and deployment of other large scale technologies.

Feminist empiricism holds as indisputable that science and scientific assumptions are shaped by history. According to Crasnow, examining scientific norms, assumptions, evidence, and theory in a context dependent way makes the role of history more visible since norms vary "from one historical period to another, from one culture to another, and, within a culture, from one group to another" (Crasnow 2003, 132). She cites Longino who affirms that, "States of affairs are taken as evidence in light of regularities discovered, believed, or assumed to hold. The evidential relations into which a given state of affairs can enter will be as varied as the beliefs about is relations with other states…" (Longino 1990, 41). As such, Longino maintains that there must be a balance between openness and truth for which evidential assessment in a social context lends robustness to science.

At its core, what is required for mutuality of interaction in climate geoengineering to be realized is recognizing that an object or set of processes (i.e. the climate system) is not static or zero-sum but is influenced by plurality of different factors, forces and arguments that go beyond what is traditionally categorized as 'the scientific.'

At the margins of scientific practice, there have been some interesting examinations of complexity with respect to, for example, the process of public opinion formation which is uniquely placed to guide geoengineering research and representation in a more processual direction. Discursively, Macnaghten et al. argue that public attitudes towards technologies like geoengineering can and should be understood by "focusing on the situated and interactive quality of public reasoning" (Macnaghten et al. 2015, 3), which is shaped by "the signature of novel technologies themselves" – i.e. "the specific ways in which [their] material features are articulated in practical reasoning and discourse within real-world settings" (Horlick-Jones 2007, 85).

Macnaghten et al. explain how emerging technologies like geoengineering are situated socio-culturally by drawing on existing research that coalesces around the themes of purpose, trustworthiness, inclusion and agency, speed and direction of innovation, and equity in terms of social benefit (Macnaghten et al. 2015, 8). These overlap with Longino's virtues and contribute to the understanding of climate engineering by setting boundaries around the research and shaping the course of its development. Macnaghten et al. also draw on narrative analysis and social theory to demonstrate the coextensive relationship between public imaginings and the scientific process. Public narratives like the "slippery slope" frame, which asserts that technological advances that "seem beneficial now will inevitably evoke further technological steps and applications that are morally doubtful," and "the 'colonization' narrative, i.e. that technology will spread out and ultimately colonize autonomy and agency" (Macnaghten et al. 2015, 12), apply to geoengineering as well.

Overall, it is difficult to overcome the barriers associated with entrenched scientific practices in order to institute methodologies and approaches that incorporate complexity, interaction, perspective and scale in ways that, as Kourany contends, steer theorizing away from simple-dominant subordinate conceptions of nature that naturalize social domination as well as linearity, to ones that are complex, interactive and mutually constitutive (Kourany 2010, 54). As demonstrated, this approach has, to a degree, been incorporated into sulphate geoengineering research with respect to the inclusion of new and dynamic modeling techniques that include the integration of feedback loops, the use of field and networked approaches, the application of pattern scaling, and the highlighting of the complexity of specific climate phenomena. Yet even these epistemological moves do not take full advantage of the kinds of mutuality and dynamism that an interactionist approach is meant to make possible. As Longino makes clear, "feminist scientists have taken complex interaction as a fundamental principle of explanation" (Longino 1995, 388). In not meeting the criteria of mutuality of interaction, climate geoengineering cannot claim to represent a scientific practice that can make inequalities, whether gendered or other, visible. If we support the assumption that Longino's theoretical virtues are essential in guiding a mode of science that is robust, objective, prosocial, egalitarian, and inclusive, criticism of climate engineering from the perspective of feminism is essential. In the next Chapter, the diffusion of power virtue is discussed with particular attention paid to how asymmetries in scientific knowledge, practice, and decision making are made manifest with respect to geoengineering.

References

Alaimo, S. (2008). Trans-corporeal feminisms and the ethical space of nature. In S. Alaimo & S. Hekman (Eds.), *Material feminisms*. Bloomington: Indiana University Press.

Andrews, T., et al. (2012). Forcing, feedbacks and climate sensitivity in CMIP5 coupled atmosphere-ocean climate models. *Geophysical Research Letters, 39*, 9.

Appelbaum, P. (2016). Mathematics education as a matter of curriculum. In M. A. Peters (Ed.), *Encyclopedia of educational philosophy and theory* (pp. 1–6). Singapore: Springer.

Barad, K. (2007). *Meeting the universe halfway: Quantum physics and the entanglement of matter and meaning.* Durham: Duke University Press.

Barad, K. (2012). On touching—The inhuman that therefore I am. *differences, 23*(3), 206–223.

Bjornerud, M. (1997). Gaia: Gender and scientific representations of the Earth. *NWSA Journal, 9*(3), 89–106.

Bleier, R. (1986). Introduction. In R. Bleier (Ed.), *Feminist approaches to science* (pp. 15–16). New York: Pergamon.

Bug, A. (2003). Has feminism changed physics? *Signs: Journal of Women in Culture and Society, 28*(3), 881–899.

Cass, L. R., & Pettenger, M. E. (2007). Conclusion: The constructions of climate change. In Kirton et al. (Eds.), *The social construction of climate change: Power, knowledge, norms, discourses* (pp. 235–246). New York: Ashgate Publishing.

Crasnow, S. (2003). In C. L. Pinnick, N. Koertge, & R. F. Almeder (Eds.), *Scrutinizing feminist epistemology: An examination of gender in science* (pp. 130–141). New Brunswick: Rutgers University Press.

Dimen, M. (1989). Power, sexuality and intimacy. In A. Jaggar & S. Bordo (Eds.), *Gender/Body/ Knowledge. Feminist reconstructions of being and knowing* (pp. 34–51). New Brunswick: Rutgers University Press.

GCRP. (2009). *Climate literacy: The essential principles of climate science.* Available at: https:// cpo.noaa.gov/sites/cpo/Documents/pdf/ClimateLiteracyPoster-8_5x11_Final4-11LR.pdf. Accessed 2 Oct 2018.

Gutiérrez, R. (2012). Embracing "Nepantla": Rethinking knowledge and its use in teaching. *REDIMAT-Journal of Research in Mathematics Education, 1*(1), 29–56.

Heckman, S. (1990). *Gender and knowledge.* Boston: Northeastern University Press.

Heyward, C., & Rayner, S. (2013). *A curious asymmetry: Social science expertise and geoengineering.* Climate Geoengineering Governance Project Working Paper, 7.

Horlick-Jones, T. (2007). On the signature of new technologies: Sociality, materiality and practical reasoning. In R. Flynn & P. Bellaby (Eds.), *Risk and the public acceptance of new technologies* (pp. 41–65). Basingstoke: Palgrave Macmillan.

Horton, J. B. (2015). The emergency framing of solar geoengineering: Time for a different approach. *The Anthropocene Review, 2*(2), 147–151.

IPCC. (2007). *Fourth assessment report: Climate change 2007 (AR4). IPCC.* Available at: http:// www.ipcc.ch/publications_and_data/publications_and_data_reports.shtml#1. Accessed 15 Aug 2018.

Irvine, P. J., Ridgwell, A., & Lunt, D. J. (2010). Assessing the regional disparities in geoengineering impacts. *Geophysical Research Letters, 37*(18), L18702. https://doi.org/10.1029/2010GL044447. Accessed 21 Jan 2017.

Jones, A., et al. (2001). Indirect sulphate aerosol forcing in a climate model with an interactive sulphur cycle. *Journal of Geophysical Research: Atmospheres, 106*(D17), 20293–20310.

Keller, E. F. (1987). Feminism and science. In S. G. Harding & J. F. O'Barr (Eds.), *Sex and scientific inquiry* (pp. 233–246). Chicago: University of Chicago Press.

Kirton, J. J., et al. (2013). *The social construction of climate change: Power, knowledge, norms, discourses.* New York: Ashgate Publishing.

Kourany, J. A. (2010). *Philosophy of science after feminism.* Oxford: Oxford Univerity Press.

Kravitz, B., et al. (2016). Geoengineering as a design problem. *Earth System Dynamics, 7*(2), 469–497.

Kuebbeler, M. et al. (2012). Effects of stratospheric sulfate aerosol geo-engineering on cirrus clouds. *Geophysical Research Letters, 39*(23). https://doi.org/10.1029/2012GL053797. Accessed 22 Jan 2012.

Lahsen, M. (2002). *Brazilian climate epistemers' multiple epistemes: An exploration of shared meaning, diverse identities and geopolitics in Global Change Science.* Harvard University. Global Environmental Assesment Project.

Longino, H. E. (1990). *Science as social knowledge: Values and objectivity in scientific inquiry.* Princeton: Princeton University Press.

Longino, H. (1995). Gender, politics and the theoretical virtues. *Synthese, 104*(3), 383–397.

Longino, H. E. (1996). Cognitive and non-cognitive values in science: Rethinking the dichotomy. In L. H. Nelson & J. Nelson (Eds.), *Feminism, science, and the philosophy of science* (pp. 39–58). Dordrecht: Kluwer Academic.

Longino, H. E., & Lennon, K. (1997). Feminist epistemology as a local epistemology. *Proceedings of the Aristotelian Society*, Supplementary Volumes, *71*, 19–54.

Macnaghten, P., Davies, S. R., & Kearnes, M. (2015). Understanding public responses to emerging technologies: A narrative approach. *Journal of Environmental Policy & Planning, 10*, 1–9.

Neimanis, A., & Walker, R. L. (2014). Weathering: Climate change and the "thick time" of transcorporeality. *Hypatia, 29*(3), 558–575.

Rovelli, C. (1996). Relational quantum mechanics. *International Journal of Theoretical Physics, 35*(8), 1637–1658.

Sinclair, N. (2009). Aesthetics as a liberating force in mathematics education? *ZDM Mathematics Education, 41*(3), 45–60.

Spanier, B. (1995). *Im/partial science: Gender ideology in molecular biology.* Bloomington: Indiana University Press.

Stehr, N., & Von Storch, H. (1995). The social construct of climate and climate change. *Climate Research, 5*(2), 99–105.

Steinhaeuser, K., et al. (2011). Complex networks as a unified framework for descriptive analysis and predictive modeling in climate science. *Statistical Analysis and Data Mining, 4*(5), 497–511.

Subramaniam, B. (2014). *Ghost stories for Darwin: The science of variation and the politics of diversity.* Champaign: University of Illinois Press.

Tillman, R. (2015). Toward a new materialism: Matter as dynamic. *Minding Nature, 8*, 1. http://www.humansandnature.org/toward-a-new-materialism-matter-as-dynamic. Accessed 27 Jan 2017.

Tobias, S. M., & Marston, J. B. (2013). Direct statistical simulation of out-of-equilibrium jets. *Physical review letters, 110*(10), 104502.

Tuana, N. (2001). Material locations: An interactionist alternative to realism/social constructivism. In *Engendering Rationalities* (pp. 221–243). Albany: SUNY Press.

Wehling, P. (2006). The situated materiality of scientific practices: Postconstructivism–a new theoretical perspective in science studies? *Science, Technology & Innovation Studies, 1*(1), 81.

Whitten, B. L. (1996). What physics is fundamental physics? Feminist implications of physicists' debate over the superconducting supercollider. *NWSA Journal, 8*(2), 1–16.

Williams, J. W., Stephen, T., & Jackson, S. T. (2007). Novel climates, no-analog communities, and ecological surprises. *Frontiers in Ecology and the Environment, 5*(9), 475–482.

Yousefi, B. (1998). *Dissecting the ethical scientist: Baha'i and feminist perspectives.* MA thesis, Simon Fraser University.

Chapter 7
Diffusion of Power

Abstract This Chapter outlines the various characteristics of Helen Longino's diffusion of power virtue using concrete examples. It is maintained that the arguments that structure, assumptions, and theories that underpin sulphate SRM climate engineering is especially incongruous with the value of the diffusion of power. In addition to the power inequities related to participation in the of practice science, it is also noted that this virtue is intimately tied to the question of consequences, effects, and outcomes. As such, it is argued that what required is an analysis that includes the effects this technology will have on people, the natural world, systems of governance, and future generations.

Keywords Power · Geoengineering · Helen Longino · Representation · Haida · Iron fertilization · Anthropocene

Diffusion of power is the first two pragmatic, as opposed to epistemic, feminist virtues Longino maintains should play a role in scientific practice whether it is in relation to theory construction, research programs and processes, objectives, or decision-making. Both diffusion of power and the realization of human needs are pragmatic in the sense that they focus on scientific practices that shape theoretical understandings and have material, real-world consequences. In doing so, these virtues explicitly seek, "recognition of the technologically driven nature of science and call for certain technological infrastructure and outcomes over others." Social responsibility, according to Longino, is reflected in these virtues (Longino 2008, 76).

In this Chapter, after articulating the various characteristics of the diffusion of power virtue using concrete examples, sulphate geoengineering is assessed with these attributes and objectives in mind. I review the argument that the structure, assumptions, and theories that underpin this particular form of SRM climate engineering is especially incongruous with the values diffusion of power, but expand the argument beyond the science into the domain of policy. In addition to the study of power inequities related to participation in the practice science, it is important to note that, this virtue is intimately tied to the question of consequences, effects, and outcomes. As such, what is also required is an analysis that

T. Sikka, *Climate Technology, Gender, and Justice*, SpringerBriefs in Sociology,
https://doi.org/10.1007/978-3-030-01147-5_7

includes the effect this technology will have on people, the natural world, systems of governance, and future generations.

First, diffusion of power endorses those theories, frameworks and research programs whose cost of participation "do not require arcane expertise, expensive equipment, or that otherwise limit access to utilization and participation" (Longino 1996, 48). This is meant specifically to remove the cognitive obstacles groups traditionally excluded from scientific practice have faced that make it difficult to participate in a meaningful way. This participation, to be clear, is meant to facilitate science that is inclusive and equalizing by, for example, encouraging production of "procedures that empower individual women to make decisions about her health or to retain control over her body" (Longino 1996, 48–49).

The majority of examples Longino gives in her work stem from the biological sciences and generally attend to "the neglect of women's distinctive health issues," particularly around reproduction, but also health problems that are more intersectional as in the case of heart disease. A prescient example not explicitly discussed by Longino but particularly relevant, can be seen in the domain of cardiovascular health. It has been established that, what we take to be 'traditional' heart attack symptoms, i.e. chest pains and numbness, are less pronounced in women for whom heart attacks tend to present differently, e.g. with sharp back pain, fatigue, and shortness of breath (Barouch 2016). In 2016, a study by the University of Leeds of 6000,000 patients at 240 UK hospitals over the course of 9 years found that women had a 50% higher chance of misdiagnosis when suffering a heart attack as compared to men (Wu et al. 2016). This disparity is indicative of a significant power differential embedded in the practice of medicine that a more open, distributed, and diffuse sense of authority over research agendas, questions, and objectives might challenge.

Diffusion of power also extends to the use of distorted metaphors, as in the case of "the cultural identification of the male with activity and the female with passivity" in cell and microbiology, and the privileging of models and theories that are "constructed around relations of unidirectional control" (Longino 1993, 104). It is essential, therefore, to be cognizant of language that might impede the decentralization of power and influence which, in the environmental domain, includes similar operating metaphors like 'active humanity/passive earth' which fosters the belief that interfering with the earth's processes would be, if not benign, at least relatively straightforward. Finally, according to Wylie, decentralization of power also involves challenging accepted epistemic authority, i.e. who gets to set the agenda and produce knowledge, in order to "be assured of ongoing, rigorously critical engagement" (Wylie, 352, 1995). In the context of SRM climate engineering, those producing knowledge and making decisions tend to be experts, policy makers, and, as demonstrated below, corporate actors rather than the general public.

On the subject of participation and gender, as noted earlier, there is a significant lack of women involved in the study climate engineering in particular, but also in climate science more generally. This inequality extends beyond gender to racialized and ethnic minorities as well (UN Women and Mary Robinson Foundation 2013; National Science Foundation 2015; Sachs 2014). Having a voice at the table is an indispensable component of diffusion of power since, without it, the open, participatory and democratic norms needed to govern a progressive and epistemically robust scientific practice is cannot be attained. As Longino argues, a theory,

> ...which is the product of the most inclusive scientific community is better, other things being equal, than that which is the product of the most exclusive. It is better not as measured against some independently accessible reality but better as measured against the cognitive needs of a genuinely democratic community (Longino 1990, 214).

Having discussed the epistemic disadvantages of a lack of representativeness in detail, in addition to detailing the specific reasons why there is a lack of female climate scientists (e.g. socialization, a hostile work environment, overt discrimination, the competitive nature the of work itself, work/home balance etc.), issues that go beyond this lack of representation can be examined in more detail. This is particularly important since diffusion of power is a pragmatic virtue, which means that its analytical remit is significantly broader than the science itself. In addition to scientific practice, policy, law, and decision-making processes at all levels are also relevant to the discussion.

To begin with, it is the case that if we hold that science is value laden and that social interests and human needs drive the structure and content of science from the questions asked, to the theory formed, and the uses to which it is put, not having women at the table will inevitably result in the erasure of a significant and potentially transformational set of values, interests, questions, and concerns. In relation to geoengineering, this lack of representation is visually obvious. With the exception of a handful of scientists, including Margaret Leinen and Debra Wesenstein, women are not represented. As such, the voices, perspectives, ideas, and values of a significant portion of the population are excluded at the point of admission – i.e. the table, or in this case the lab, itself. This is a clear violation of the diffusion of power virtue as well as indicative of scientific outcomes that are likely to be less heterogenous, less novel and potentially less empirically sound.

One particular case in which the diffusion of power virtue was breached is in relation to a climate engineering test that did not involve SO_2 geoengineering, but a form of CDR mentioned previously, namely, iron fertilization. To reiterate, this particular form of geoengineering consists o firon being placed into the ocean in order to increase plankton blooms that would sequester carbon. In 2012, Russ George, an American businessman and geoengineering proponent, partnered with a the Haida indigenous community in Northern British Columbia (Old Massett) to dump 100 tons of iron sulphate into the ocean ostensibly to increase salmon stocks and serve as a source of revenue in the form of purchasable carbon credits. After the project was executed, it was roundly condemned by NGO's, activists, scientists and governments including a pointed condemnation by signatories to the Convention on Biological Diversity for which the underlying convention and protocol (the London Convention and Protocol respectively) restricts the dumping of substances in the sea. Resolutions to clarify and further restrain such actions, consistent with the precautionary approach, have been made, yet most are are non-binding and leave open the possibility of continued research (Craik et al. 2013). There is an ongoing investigation of this by the Canadian Environmental Agency and the RCMP who, in 2014, executed search warrants on the corporate offices of the two companies that produced the iron and supplied the ship from which the iron was dumped. After some legal back and forth, the investigation continues with no formal charges filed (Moore 2013). It should be noted that although George, the Haida Community, and the company they created to conduct this trial all deny that this was a geoengineering experiment, George's ownership of Planktos, a now defunct company aimed precisely at ocean restoration and *slowing climate change*, as well as his previous attempts at this form of geoengineering, point in the opposite direction. The investigations themselves are operating on the assumption that this was in fact a geoengineering trial under the guise of salmon restoration.

In this case, climate engineering represents the apex of postnormal science manifest in techniques with a particularly complicated relationship to power, control, and authority. Most significantly, the traditional view of Western science dominating indigenous knowledge is also challenged since the Old Massett Council, representing the Haida Nation, were joint partners with George's company in this project. As Amy Hinterberger points out, the company established jointly between the two parties, called the Haida Salmon Restoration Corporation, played up the joint relationship between the two groups which they promoted as the central "idea behind the action: 'World class science, village style'" (Hinterberger 2013). While it could be argued that, this is an instance of a traditionally marginalized community partnering, by choice, with a company with joint objectives, others have pointed out that colonialism is indeed part of the narrative and that the community's high unemployment and dwindling reserves of salmon placed the community in a particularly vulnerable position. As such, there is a clear lack of parity between parties and, as such, an unmistakable infringement on the diffusion of power. Part of the significance of this case is its precedent setting role in mapping out how geoengineering might unfold. Questions around who or what nation/body makes the decision to geoengineer, where geoengineering takes place, who is negatively affected, what is the nature of the consent given, what are the rules around compensation, and who decides when to pull the plug etc., exposes a raft of instances in which power will likely be unequally distributed since, in most cases, it is people with political and economic power who will be the ultimate decision makers.

This brings us to the role of expert knowledge which is significant in this context since it is the scientists themselves who work to legitimate science, justify technological advance, and provide the information on which decisions are made (Jasanoff 2004). Yet, in the case of climate engineering, at each stage the public has tended to be been left out of the conversation. This erasure is not unusual when the underlying science is complex, difficult, and often at odds with itself as in the cases of nanotechnology, some forms of biotechnology, and astronomy as well. This latter point is significant since it is also true that, in a handful of other cases, the public, or at least interested segments of it, have successfully intervened to guide scientific practice.

Before elaborating on one such instance, it is important to note that in the case of geoengineering, the public at large is spatially removed from the science since the majority of the modeling and visualizations take place at large universities or corporate laboratories in closed settings. In the case of laboratory research in particular, it is difficult to imagine how public participation, specifically with respect to modeling and simulation, would look like. Having discussed how public participation could be fostered in earlier Chapters, particularly with respect to decision-making about its potential use, and having reached the conclusion that citizen engagement is considerably easier to envision on the deployment and policy side (Parkhill et al. 2013; Pidgeon et al. 2013; Carr et al. 2013), it would be beneficial to spend a little time examining what successful citizen engagement in science might look like. A prime example comes from citizen engagement in HIV/AIDS research, treatment, and trials in the 1990s. In this case, motivated activists, many from the gay community, pushed for changes in the conservative FDA and NIH (Food and Drug Administration and National Institutes of Health respectively) policies around experimental treatments and for an increase in medical trials that were oriented to pragmatic questions of effective treatment. In order to gain legitimacy, these activists began to acquire "the language and culture of medical science" by "attending scientific conferences, scrutinizing research protocols" and, in doing so, gaining "a working knowledge of the medical vocabulary" (Epstein 1995, 417).

After significant organizing, activists were indeed invited to the table and their gained expertise and perspectives led to their inclusion on review committees that allocated funding and accessed the science; increased public participation at conferences; began to study HIV related conditions associated with women specifically; and instituted new regulatory and interpretive mechanisms (Epstein 1995; Barr et al. 1992; Corea 1992; Edgar and Rothman 1990). The success of these groups, mostly located in NYC and the San Francisco Bay area, is a key example of the democratization of science where decision making power was ultimately decentralized with even scientists remarking on the positive contributions made by citizens to the practice of science itself (Epstein 1995).

Activists in climate science have been similarly engaged in mobilizing around climate change with global groups like 350.org, The Sierra Club, Idle No More, and Greenpeace working with local groups and affiliates that have taken on advocacy roles and some research as well (the majority of these groups have climate scientists as members). Climate science has made some movements towards crowd-sourcing data collection and smaller scientific tasks, including ecological data, such as ground weather tracking, and the tracking of pollution levels, water quality, species numbers aided by technological advances in GIS web applications, data informatics and even smart phones (Dickinson et al. 2010; Root et al. 2003; Field et al. 2003). There are a host of organizations, involved in such work including Citizen Science Central, the Data Observation Network for Earth, and SciStarter. It has been argued that this kind of citizen engagement in scientific practice promotes a sense of stewardship amongst communities that is necessary to tackle climate change (Chapin et al. 2011). However, it is not the case that a truly collaborative science, where problem definition, data collection and analysis, often called 'extreme citizen science,' is open to those outside the academic community, has been realized in relation to climate engineering (Haklay 2012; Science Communication Unit 2013).

It is significant that diffusion of power expressed through active engagement and participation *in the practice of science* is not something widely discussed or researched in the literature. There is, however, some compelling work being done at the levels of advocacy for and against geoengineering as well as research on policy formation and public perceptions (Corner et al. 2013; Poumadère et al. 2011; Carr et al. 2013). SO_2 climate engineering research, in which "lay citizens challenge the rules of the scientific method and are involved in the production of knowledge" (Kleinman 2000, 141), are virtually non-existent with, at this current juncture, little in the way of openings or opportunities. It could be argued that the rise in private organizations and actors outside academe that have become involved in geoengineering research represent a kind of diffusion of power– as in the case of Russ George's company – but this is not the kind of participation consistent with Longino's vision which decentralizes-downwards and would demand the use of less control-oriented technologies like,

> …intensive small-scale sustainable agriculture, promoting health by preventive measures such as improved hygiene rather than high-tech interventive measures available only to the few, protection of the environment by conservation and widely dispersed renewable energy technologies (Longino 1995, 388).

Understood in this way, geoengineering technologies like sulphate SRM can be seen as structurally inimical to a diffusion of power since, by their very nature, these technologies require institutions and a scientific culture that are hierarchically designed to limit risks and effectively act as a veritable thermostat through time and space. I return to this issue of hierarchy and control further on in a discussion of the Anthropocene.

Following from this difficulty, and remaining on the subject of scientific practice, Longino makes clear that, science that is equalizing should not only involve diverse actors, but also ensure that technically interventionist measures are not "measure(s) available only to the few" (Longino 1996, 48–49). This raises questions around access and specifically on what geoengineering 'available for use by the many' would look like since it, by definition, it is a technoscientific endeavor that is large scale and ecologically transformative. It therefore requires centralized and exclusive decision-making by definition. The notion that geoengineering, with respect to values, demands a hierarchical and top down command structure speaks directly to its lack of feminist bona fides. Yet it remains the case that the risk of multiple Russ George type actions would not be desirable, safe, or efficacious leading to the conclusion that perhaps open access to these technologies, in the manner articulated by Longino, should be approached carefully.

Building on this argument, as Bronson contends, it is important to underscore that any notion of small-scale geoengineering is in fact an oxymoron since it is a technology that requires large scale deployment in "order to yield the critical information to become a reliable technology" (Bronson 2011, 38). Similarly, Factor points out that even in the case of field trials, all forms of geoengineering necessitates a level of deliberate deployment which, consequently, means that they "might better be conceived [of] as a 'real-world experiment[s]'" (Factor 2015, 314; Krohn and Weyer 1994).

As Alan Robock makes clear, SO_2 geoengineering in particular can only be implemented, or even adequately tested, on a large-scale. Even decisions on particle size must consider how aerosol droplets,

> ...will grow in size (which determines the injection needed to produce a particular cooling) [which] can only be tested by injection into an existing aerosol cloud, which cannot be confined to one location. Furthermore, weather and climate variability preclude observation of the climate response without a large, decade long forcing (Robock et al. 2010, 530–531).

Thus, participation is geographically and temporally restricted in terms of decision-making, testing, and deployment. This inevitably restricts civil society engagement at all levels of application to say nothing of the practice science itself.

It is also important, in relation to the diffusion of power, to consider the precise mechanisms of decision-making through which consensus formation actually occurs. It is here that Longino's conception of science as a social practice makes another appearance. Decentralization of power holds that critical and inclusive engagement in all aspects of science be maintained and equality of intellectual authority preserved (Longino 2002). Considered in this way, and in highlighting the twin objectives of access and participation, Longino's diffusion of power virtue is again prevented from being realized as it relates to climate geoengineering. Having established that substantive engagement in the practice of science is itself lacking, except for post-facto consultation on policy and questions of implementation, it is also important to highlight the considerable gap between lay and expert knowledge.

This is particularly evident, with respect to the domain of climate science, which makes it difficult for any degree of equality of intellectual authority to be realized. However, it also the case that the type and kind of equality Longino is interested in has to do more with the basic capacity to engage in critical discussion, debate, and response as the conditions of participation, rather than expert knowledge. When parties representing different perspectives are seen as authoritative in this way, more room is made for alternative standpoints that have to be defended to gain cognitive status. It is more likely, as a result, that marginalizing practices are challenged and that the contributions of women and minorities are highlighted leading to new spaces for the production of knowledge that does "not naturalize relations of domination but that offers other ways to interact with the natural world, and by extension, with one another" (Longino 2008, 84).

Accordingly, it is not necessary for all participants to be able to understand the intricacies of an AOGCM model, but that their perspectives, whatever they are, are respected and given due consideration so that, when consensus does exists, it "must not be the result of the exercise of political or economic power or of the exclusion of dissenting perspectives; it must be the result of critical dialogue in which all relevant perspectives are represented"(Longino 1993, 113). This is a compulsory component of diffusion of power since, without it, dissenting voices would either be disregarded outright or assigned token consideration. It also lessens the likelihood that novel questions, alternative ideas and hypotheses, and innovative perspectives are ignored – particularly from communities that lack social, economic or political power. As Wynne argues, public engagement with science and scientists "represents an iterative process between lay people and technical experts rather than a narrowly didactic or one-way transmission of information packages" (Wynne 1991, 114). It can also facilitate a useful degree of social learning as well as affording an outlet through which to exercise citizenship rights (Parkhill et al. 2013, 220). The actualization of this virtue would also ensure that values, standards and beliefs are held to account. This might include an interactive rather than hierarchical, local rather than global, and differentiated rather than singular approach to climate science and geoengineering.

As has been demonstrated, the assumptions and values constitutive of science shape outcomes, beliefs and scientific possibilities. At present, it is not apparent that climate engineering has made strides towards realizing the diffusion of power virtue. This is the case not only because geoengineering research is asymmetrically organized and geographically concentrated, with experts in universities and private actors/institutions/think tanks playing a dominant role, but also because those experts and groups are themselves not representative or diverse in any meaningful way. It is at this point that the lack of diversity becomes an epistemic problem.

This brings us to the topic of the Anthropocene – which has been intentionally left out of the discussion until this point because of its specific relevance to FCE's decentralization of power virtue. The Anthropocene, briefly, represents the view that changes perceived in the natural world are no longer independent of, or just partially influenced by, human activity. Rather, it holds that global changes in the Earth's geology and ecology are being driven in large part *by* humanity itself. Interestingly, Paul Crutzen, the Nobel Prize winning pioneer of both SRM and CDR geoengineering research, is credited with popularizing the term (Crutzen 2002). It is part of what he and others call the 'Earth System Model' in which the Earth is seen as being made of up of interacting systems (chemical, physical, biological, global) and cycles that support life (Zalasiewicz et al. 2011; Williams et al. 2011). It also holds that "biological/ecological processes are an integral part of the functioning of the Earth Systems," that "forcings and feedbacks within the Earth system are as important as external drivers of change," and, finally, that "the Earth system includes humans, our societies, and our activities" (Steffen et al. 2007, 615).

What is interesting about the Anthropocene, when compared to the Holocene that came before it and under which society developed through industrialization, mechanization, urbanization etc., is how it shapes our understanding of our relationship to the word – specifically with respect to whether we see ourselves as separate and apart, or an inextricable part of the natural world. The belief that humanity and nature exist in distinct ontological zones where we, as coherent, autonomous, and self-possessed persons, are entitled to ownership and autonomy of action over nature, is an outcome of distinctly Western, European, and modernist conception of the market, politics, culture, and ethics. As Whatmore argues, the agency extended to humans is "predominantly defined in terms of social (human) relations. Where they are addressed at all, environmental nonhuman relations are treated as passive contextual extensions of human well-being" (Whatmore 1997, 41). It is thus possible, accordingly, to see geoengineering as a logical continuation of the challenges posed by the Anthropocene where the former is understood as extending the traditional, hierarchically structured, and exploitative human-nature relationship.

The language used in sulphate geoengineering research exemplifies the lack of diffuse power since it is based on an understanding of the human-nature relationship that fits nicely into this largely Western epistemology. Taking the GeoMIP article by Kravitz et al. (2011) as an example, where is ample use of language that takes for granted *our* decision to *implement* geoengineering on a *passive* environment, *our* modeling *of* the earth and *our* experiments *on* it which assumes a passive natural world lacking agency. As such, phrasing that includes statements like the following,

> These idealized experiments are expected to reveal the basic model responses to this forcing balance without the added complication of differing treatments of stratospheric aerosols in the various models (Kravitz et al. 2011, 163)

...places power in the hands of scientists whose experiments act on the world through the imposition of their idealizations, their chosen levels of forcing, their derived and applied equations, and thier reliance on values and assumptions made without discussion such as acceptable levels of warming, levels of GHG emission, disruption to monsoons, and effects on ozone chemistry (Kravitz et al. 2011). Power is thus exercised over nature which is scientifically constructed as an object to be studied which, as Swyngedouw maintains, is not only "simplistic, reductionist, teleological and, ultimately, homogenizing," and assumes that there is one nature or environment that is both "trans-historical and trans-geographical" (Swyngedouw 2011). It is on this latter point that sulphate geoengineering falls short. Research aims to "investigate how SRM might be used to counterbalance future GHG-induced climate change," through a set of research protocols that "simulate past and future climate scenarios using a wide range of model parameter combinations that both reproduce past climate within a specified level of accuracy" (Ricke et al. 2011) assumes, implicitly, that nature has no, or at the least very little, authority. For Longino's diffusion of power virtue to be realized, it would require that the natural world be empowered as an active agent with its own moral status – thus moving us away from an instrumental view of the environment.

Addressing this in a manner consistent with Longino's virtue might include the adoption of an ecocentric, holistic and intuitive view of nature as held by conservationists (e.g. the deep ecology movement) (Bookchin 1987; Lynch and Norris 2016) or even an ecofeminist rejection of what is seen as the patriarchal domination of nature for a more co-equal and integrated relationship that draws on metaphors like the web or networks to symbolize "the interdependence that is thought to characterize human relations, according to feminist thought" (Lauritzen 1989, 33–34) (Warren 1990; King 1991). Translating and applying the possible instantiations of this virtue is difficult primarily because it contradicts the entire ethos, logic, and rational for SRM climate engineering. As such, deep ecology and ecofeminism have little to offer by way of nuanced critique, other than rejection, as well as having insuperable epistemological and ontological disagreements with feminist empiricism on the subjects of truth, knowledge and reason that makes it difficult to rely on them as alternatives (Harding 1986a, b; Merchant 1994; Zimmerman 1994; McDowell 1993, 2016; Intemann 2010). That being said, Longino does commend "ecofeminists and feminists in developing regions" who, "urge the development of technologies that are accessible and that can be locally implemented" (Longino 1995, 389). However the complications posed by localized geoengineering, with limited access to technologies that would allow this, is both difficult and, because of its inescapably global effects, potentially dangerous.

One interesting way in which nature's status can be productively challenged in relation to climate science and geoengineering (thereby distributing power more equitably), is in recent moves to confer legal status onto nature itself. This approach is rooted in the doctrine of the public trust, which confers responsibilities onto the State for safeguarding the natural world for future generations. Recently, legal challenges have been launched in order for the environment to have legal standing before the law and, in 2008, Ecuador made changes to its constitution such that it now recognizes that,

> …the inalienable rights of ecosystems, give individuals the authority to petition on the behalf of ecosystems, and require the government to remedy violations of nature's rights, including the right to exist, persist, maintain and regenerate its vital cycles, structure, functions and its processes in evolution (Constitution of the Republic of Ecuador 2008; Shelton 2015).

These changes also empower individuals and communities to act on behalf of nature to enforce stated principles through legal constitutional procedures and lays out the rights 'nature' has to damages in the case of "extinction of species, devastation of ecosystems or permanent change of natural cycles" (Lalander 2014, 158; Case Note 2012).

There is also interesting advocacy work being done by groups like Our Children's Trust, which is a group made up of 21 youth plaintiffs who have launched a constitutional lawsuit against the US government claiming that "through the governments affirmative actions in causing climate change, it has violated the youngest generation's constitutional rights to life, liberty, property, as well as failed to protect essential public trust resources" (Our Children's Trust 2017). While the Our Children's Trust example does not go as far as the Ecuador case in redefining our relationship to nature in a decentralized way, it does introduce a challenge to geoengineering research which, because it is hierarchically structured, still conceives of nature as something to be acted on by 'big science,' and is thus resistant to citizen participation. It is challenging to envision how geoengineering might be litigated if tests result in environmental damage or, if deployed, how affected parties might go about seeking compensation and damages not only for material losses but also for emotion and spiritual suffering (Horton et al. 2014; Reynolds 2014). It is also unlikely that litigating a diffusion of power is the most productive path forward. I return to the issues of law, litigation, and regulation in the next Chapter.

Finally, it is also the case that, with respect geoengineering, diffusion of power must be assessed in the context of potential commercialization. The move towards a decentralization of power should not, within a feminist framework, results in the empowerment of commercial forces at the expense of the public good which, in this instance, is a concern. Key examples include the case of Russ George and his unsanctioned climate engineering test as well as billionaires like Bill Gates and Richard Branson who are funding scientists engaged in geoengineering research – ostensibly for honorable ends, but also for the inevitable patents, profits, and control it would produce (Reynolds 2011; Yusoff 2013; Sikka 2013). This will result in power being centralized in the hands of private interests rather than being dispersed throughout the polity. Longino is clear that her virtues should favor frameworks that "reduces [sic] inequalities of power," and which "requires taking stock of the... economic context on which development might take place" (Longino 1995, 394).

Taken as a whole, Longino's pragmatic diffusion of power virtue articulates an ideal of intellectual practice aimed at ensuring that traditional hierarchies and concentrations of power are prevented from distorting knowledge and simultaneously guaranteeing that the science we do get is consistent with feminist values. From issues of participation, metaphors, and rogue tests, to considerations of expert knowledge, activism, the Anthropocene and the legal status of nature, this particular virtue is one with concrete epistemological, ontological and practical implications for how we see the world and how we use science and technology. Diffusion of power is the condition on which a rigorous but value laden science can be challenged such that values that are exclusively subjective or distorted in some way are weeded out through critical discourse. As such, and in line with its feminist ethos, this virtue ensures that reason exercised in a social context produces science that rejects values preoccupied by hierarchy and control, e.g. with respect to the way which geoengineering research is conducted, for one that seeks scientific norms and values that are exercised in a context of openness and consensus-formation. Unjust social biases and the centering of gender are central to the FCE approach. As it stands, the ability for SRM geoengineering (comprised of the technology itself, the science behind it, and the existing frameworks) to enhance participation beyond the obligatory consultations on use and policy would require considerable structural change. Again, the values that would be required, were climate engineering to take this up, include theoretical pluralism, an open accounting of scientific norms, and agreement that "scientific judgment should [always and consistently] be held to evidential standards" (Longino 2002, 573). Longino ascribes to diffusion of power the ability to fulfill the objective of further democratization in pursuit a feminist scientific practice that creates a more symmetrical relationship between what is referred to as "the aims of individuals engaged in scientific inquiry," and "those of the society that supports such inquiry" (Longino 1988, 574). Next we take up the final virtue, human needs, before discussing some of the challenges posed to Longino's approach and the ways in which feminist standpoint theory and the technofeminist approach might be helpful – particularly in rounding out a feminist critique of geoengineering.

References

Barouch, L. (2016). *Heart disease: Differences in men and women*. John Hopkins Medicine, Heart & Vascular Institute. http://www.hopkinsmedicine.org/heart_vascular_institute/clinical_services/centers_excellence/womens_cardiovascular_health_center/patient_information/health_topics/heart_disease_gender_differences.html. Accessed 1 Feb 2017.

Barr, et al. (1992) *A model for AIDS drug development*. Presentation at Eighth International Conference on AIDS, Amsterdam.

Bookchin, M. (1987). *Social ecology versus deep ecology: A challenge for the ecology movement* (pp. 4–5). Green Perspectives: Newsletter of the Green Program Project.

Bronson, D. (2011). *Earth grab: Geopiracy, the new biomassters and capturing climate genes*. Nairobi/New York: Fahamu/Pambazuka.

Carr, W. A., et al. (2013). Public engagement on solar radiation management and why it needs to happen now. *Climatic Change, 121*(3), 567–577.

Case Note. (2012). The Ecuadorian exemplar: The first ever vindication of constitutional rights of nature. *Review of European Community & International Environmental Law, 21*(1), 63.

Chapin, S. F., et al. (2011). Earth Stewardship: Science for action to sustain the human-earth system. *Ecosphere, 2*(8), 1–20.

Constitution of the Republic of Ecuador. (2008). *Political Database of the Americas (PDBA)*.http://pdba.georgetown.edu/Constitutions/Ecuador/english08.html. Accessed 14 Feb 2017.

Corner, A., et al. (2013). Messing with nature? Exploring public perceptions of geoengineering in the UK. *Global Environmental Change, 23*(5), 938–947.

Correa, G. (1992). *The invisible epidemic: The story of women and AIDS*. New York: Harper Collins.

Craik, N., et al. (2013). Regulating geoengineering research through domestic environmental protection frameworks: Reflections on the recent Canadian ocean fertilization case. *Carbon & Climate Law Review, 7*, 117–124.

Crutzen, P. J. (2002). Geology of mankind: The Anthropocene. *Nature, 415*, 23.

Dickinson, J. L., Zuckerberg, B. J., & Bonter, D. N. (2010). Citizen science as an ecological research tool: Challenges and benefits. *Annual Review of Ecology, Evolution, and Systematics, 41*, 149–172.

Edgar, H., & Rothman, D. J. (1990). New rules for new drugs: The challenge of AIIDS to the regulatory process. *Milibank Quarterly, 68*, 111–142.

Epstein, S. (1995). The construction of lay expertise: AIDS activism and the forging of credibility in the reform of clinical trials. *Science, Technology & Human Values, 20*(4), 408–437.

Factor, S. (2015). The experimental economy of geoengineering. *Journal of Cultural Economy, 8*(3), 309–324.

Field, D., Voss, P., Kuczenski, T., et al. (2003). Reaffirming social landscape analysis in landscape ecology: A conceptual framework. *Society & Natural Resources, 16*, 349–361.

Haklay, M. (2012). Citizen science and volunteered geographic information – Overview and typology of participation. In D. Z. Sui, S. Elwood, & M. F. Goodchild (Eds.), *Crowdsourcing geographic knowledge: Volunteered geographic information (VGI) in theory and practice* (pp. 105–122). Berlin: Springer.

Harding, S. (1986a). *The science question in feminism*. Ithaca: Cornell University Press.

Harding, S. (1986b). The instability of the analytical categories of feminist theory. *Signs: Journal of Women in Culture and Society, 11*(4), 645–664.

Hinterberger, A. (2013). Curating the postcolonial critique. *Social Studies of Science, 43*(4), 619–628.

Horton, J. B., et al. (2014). Liability for solar geoengineering: Historical precedents, contemporary innovations, and governance possibilities. *NYU Environmental Law Journal, 22*, 225–273.

Intemann, K. (2010). 25 years of feminist empiricism and standpoint theory: Where are we now? *Hypatia, 25*(4), 778–796.

Jasanoff, S. (2004). *States of knowledge: The co-production of science and the social order.* London: Routledge.

King, R. J. (1991). Caring about nature: Feminist ethics and the environment. *Hypatia, 6*(1), 75–89.

Kleinman, D. L. (Ed.). (2000). *Science, technology, and democracy.* New York: SUNY Press.

Kravitz, B., et al. (2011). The geoengineering model intercomparison project (GeoMIP). *Atmospheric Science Letters, 12,* 162–167.

Krohn, W., & Weyer, J. (1994). Society as a laboratory: The social risks of experimental research. *Science and Public Policy, 21*(3), 173–183.

Lalander, R. (2014). Rights of nature and the indigenous peoples in Bolivia and Ecuador: A strait-jacket for progressive development politics? *Iberoamerican Journal of Development Studies, 3*(2), 148–173.

Lauritzen, P. (1989). A feminist ethic and the new romanticism – mothering as a model of moral relations. *Hypatia, 4*(3), 29–44.

Longino, H. E. (1988). Science, objectivity, and feminist values. *Feminist Studies, 14*(3), 561–574.

Longino, H. E. (1990). *Science as social knowledge: Values and objectivity in scientific inquiry.* Princeton: Princeton University Press.

Longino, H. (1993). Subjects, power, and knowledge: Description and prescription in feminist philosophies of science. In L. Alcoff & E. Potter (Eds.), *Feminist Epistemologies* (pp. 101–120). New York: Routledge.

Longino, H. E. (1995). Gender, politics, and the theoretical virtues. *Synthese, 104*(3), 383–397.

Longino, H. E. (1996). Cognitive and non-cognitive values in science: Rethinking the dichotomy. In L. H. Nelson & J. Nelson (Eds.), *Feminism, science, and the philosophy of science* (pp. 39–58). Dordrecht: Kluwer Academic.

Longino, H. (2002). Reply to Philip Kitcher. *Philosophy of Science, 69*(4), 573–577.

Longino, H. E. (2008). Values, heuristics, and the politics of knowledge. In M. Carrier, D. Howard, & J. Kourany (Eds.), *The challenge of the social and the pressure of practice, science and values revisited* (pp. 68–86). Pittsburg: University of Pittsburg Press.

Lynch, T., & Norris, S. (2016). On the enduring importance of deep ecology. *Environmental Ethics, 38*(1), 63–75.

McDowell, L. (1993). Space, place and gender relations: Part II. Identity, difference, feminist geometries and geographies. *Progress in Human Geography, 3,* 305–318.

McDowell, L. (2016). *Space, gender, knowledge: Feminist readings.* New York: Routledge.

Merchant, C. (1994). *Ecology: Key Concepts in critical theory.* Atlantic Highlands: Humanities Press.

Moore, D. (2013). Ocean fertilization experiment loses in B.C. court; charges now likely. *The Globe and Mail.* http://www.theglobeandmail.com/news/british-columbia/ocean-fertilization-experiment-loses-in-bc-court-charges-now-likely/article16672031/ 3 February 2014. Accessed 13 Feb 2017.

National Science Foundation, National Center for Science and Engineering Statistics. (2015). *Women, minorities, and persons with disabilities in science and engineering: 2015.* Arlington: National Science Foundation. Retrieved February 2, 2017.

Our Children's Trust. (2017). *Landmark U.S.* Federal climate lawsuit. https://www.ourchildrenstrust.org/us/federal-lawsuit/. Accessed 14 Feb 2017.

Parkhill, K., et al. (2013). Deliberation and responsible innovation: A geoengineering case study. In Owen et al. (Eds.), *Responsible Innovation* (pp. 219–240). London: Wiley.

Pidgeon, N., et al. (2013). Deliberating stratospheric aerosols for climate geoengineering and the SPICE project. *Nature Climate Change, 3,* 451–457.

Poumadère, M., et al. (2011). Public perceptions and governance of controversial technologies to tackle climate change: Nuclear power, carbon capture and storage, wind, and geoengineering. *Wiley Interdisciplinary Reviews: Climate Change, 2*(5), 712–727.

Reynolds, J. (2011). The regulation of climate engineering. *Law, Innovation and Technology, 3*(1), 113–136.

Reynolds, J. (2014). Climate engineering field research: The Favorable Setting of international environmental law. *Washington and Lee Journal of Energy, Climate, and the Environment, 5*(2), 417–486.

Ricke, K. L., et al. (2011). Effectiveness of stratospheric solar-radiation management as a function of climate sensitivity. *Nature Climate Change: Letters.* https://doi.org/10.1038/nclimate1328. Accessed 12 Feb 2017.

Robock, A., et al. (2010). A test for geoengineering? *Science, 327*(5965), 530–531.

Root, T. L., Price, J. T., Hall, K. R., et al. (2003). Fingerprints of global warming on wild animals and plants. *Nature, 421*, 57–60.

Sachs, C. (2014). *Women working in the environment: Resourceful natures.* Abingdon: Taylor and Francis.

Science Communication Unit, University of the West of England, Bristol. (2013). *Science for environment policy in- depth report: Environmental citizen science.* European Commission DG Environment. http://ec.europa.eu/science-environment-policy. Accessed 12 Feb 2017.

Shelton, D. (2015). Nature as a legal person. *Vertigo, 22.*https://vertigo.revues.org/16188#ftn59. Accessed 11 Feb 2017.

Sikka, T. (2013). An analysis of the connection between climate change, technological solutions and potential disaster management: The contribution of geoengineering research. In W. Filho (Ed.), *Climate change and disaster risk management* (pp. 535–551). Berlin: Springer.

Steffen, W., Crutzen, P. J., et al. (2007). The Anthropocene: Are humans now overwhelming the great forces of nature? *Ambio, Sciences Module, 36*(8), 614–621.

Swyngedouw, E. (2011). Whose environment? The end of nature, climate change and the process of post-politicization. *Ambiente & Sociedade, 14*, 2. https://doi.org/10.1590/S1414-753X2011000200006. Accessed 12 Feb 2017.

UN Women and Mary Robinson Foundation. (2013). *The full view: Advancing the goal of gender balance in multilateral and intergovernmental processes.* UN Women.

Warren, K. (1990). The power and the promise of ecological feminism. *Environmental Ethics, 12*(2), 125–146.

Whatmore, S. (1997). Dissecting the autonomous self: Hybrid cartographies for a relational ethics. *Environment and planning D: Society and Space, 15*(1), 37–53.

Williams, M., et al. (2011). The Anthropocene: A new epoch of geological time? *Philosophical Transactions of the Royal Society A., 2011*(369), 835–1111.

Wu, J., et al. (2016). Impact of initial hospital diagnosis on mortality for acute myocardial infarction: A national cohort study. *European Heart Journal: Acute Cardiovascular Care.* https://doi.org/10.1177/2048872616661693. Accessed 1 Feb 2017.

Wylie, A. (1995). Doing philosophy as a feminist: Longino on the search for a feminist philosophy. *Philosophical Topics, 23*(2), 345–358.

Wynne, B. (1991). Knowledges in context. *Science, Technology & Human Values, 15*(1), 111–121.

Yusoff, K. (2013). The geoengine: Geoengineering and the geopolitics of planetary modification. *Environment and Planning A, 45*(12), 2799–2808.

Zalasiewicz, J., et al. (2011). Stratigraphy of the Anthropocene. *Philosophical Transactions of the Royal Society A., 369*, 1036–1055.

Zimmerman, M. E. (1994). *Contesting Earth's future: Radical ecology and postmodernity.* Berkeley: University of California Press.

Chapter 8
Applicability to Human Needs

Abstract This Chapter discusses the virtue of applicability to human needs and makes that case that it has generally failed to be applied to scientific practice as well as geoengineering in particular. Issues discussed include the extent and nature of scientific backing, how ethics figures into the support of climate engineering, as well as matters related to intergenerational justice, globalization, socially responsible science and conflicts of interest. Despite the fact that a basic understanding of human needs is both intuitive and self-evident, related issues around human interests, justice, equity and power are also important and significantly complicate how this virtue is perceived and the avenues through which it can be attained.

Keywords Human needs · Globalization · Intergenerational · Feminist science · Justice · Moral hazard · Public policy · Governance

Applicability to human needs is the final feminist virtue rounding out Longino's contextual empiricist approach. It is a rather intuitive virtue that Longino spends less direct time on, but which she considers to be a central guiding principle of scientific endeavor. In this Chapter, after an explanation of the virtue itself, considerable time is spent on this how this virtue has, and has failed, to be applied to scientific practice in general as well as geoengineering. Issues discussed include the extent and nature of scientific backing, how ethics figures into the support of climate engineering, as well as matters related to intergenerational justice, globalization, socially responsible science and conflicts of interest. Despite the fact that a basic understanding of human needs is both intuitive and self-evident, related issues around human interests, justice, equity and power are also important and significantly complicate how this virtue is perceived and the avenues through which it can be attained.

As a pragmatic virtue, the objectives associated with meeting human needs include producing useful knowledge aimed at "improving the material conditions of human life, or alleviating some of its miseries" (Longino 1990, 389). Scientific programs that address hunger, poverty, deprivation, discrimination, health disparities, and environmental degradation are posited as alternatives to agendas that are exploitative, aimed at exerting power over nature or others, or results in the destruction of the environment. Longino maintains that while knowledge for knowledge's

T. Sikka, *Climate Technology, Gender, and Justice*, SpringerBriefs in Sociology, https://doi.org/10.1007/978-3-030-01147-5_8

sake remains important, scientific research must also orient itself to "knowledge and its application in technologies which empower beneficiaries" as preferable "to those which produce or reproduce dependence relations" (Longino and Lennon 1997, 23). This virtue is often contrasted with the Kuhnian values of fruitfulness and testability.

Longino also emphasizes that, science which directly meet human needs tend to be aimed at those needs traditionally ministered to by women (Longino 1996, 48). As such, it should aspire to redressing past injustices as well as setting the agenda for the way forward. Again, what is unique about Longino's approach is that, in attending to feminist interests, the commitment to producing theories that are empirically robust is retained. However, the difficulty posed by this particular approach to human needs is that, while useful analytically and as a form of generalized norm building, it is insufficiently epistemologically robust or pragmatically well defined. A useful supplement that might address this aporia can be found in Martha Nussbaum's feminist conception of human capabilities which consist of the right to life, bodily health and integrity, imagination, emotions, practical reason, affiliation, other species, play, and control over one's environment that constitute humanity's basic needs, including material essentials, as well as "diversity, pluralism and personal freedom" (Nussbaum 1999, 54, 61).

Human Capabilities: Caring Science and the Gaia Hypothesis

An interesting application of this virtue that meets both the limited principles of Longino's approach as well as the more expansive ones articulated by Nussbaum, can be seen in the study of nursing in which the pursuit of a "caring science" has gained both scientific interest and epistemological standing. The primary motive of caring science is to, "alleviate human suffering and preserve and safeguard life and health" (Eriksson 2002, 62; Lindholm and Eriksson 1993). It considers practices of caring as constituting a form of implicit and explicit knowledge that is ethical, philosophical, and epistemic (Watson and Smith 2002, 455). A caring science includes such principles as assistance with gratification of human needs, development of a human caring relationship, and the systematic use of scientific (creative) problem-solving processes that facilitates multiple ways of knowing (Watson 2007, 131).

For instance, the integration of caring science into patient care in heart health has led to a method called Quick Coherence which is a heart-centric, breathing-based, and emotional set of practices that calls for authentic presence, heart-centric emotional awareness and equanimity for self and other. Significantly, it also has explicit research protocols that focus on "the profound connection between Caring Science and scientifically based heart-centered methods" (Watson 2012). The central ethos of caring science demonstrates Longino's virtue of human needs through its guiding principles and pragmatic objective of facilitating beneficial outcomes. It is difficult to imagine climate science or geoengineering integrating such an approach, yet a move to consider caring for nature and others, a respect for relatedness and unity, as well as general equanimity can be integrated in climate research as well (Watson 2012).

In a related example of scientific practice that thematizes human needs, but which does so in relation to the natural world, Marcia Bjornerud (1997), in an article titled "Gaia: Gender and Scientific Representations of the Earth," evaluates this virtue with respect to the Gaia hypothesis which is posed as an alternative to the hierarchical, Darwinian, and interventionist tradition of Western science. Yet, instead of expounding on principles of Gaian philosophy, Bjornerud uses the Gaia metaphor to call for practical yet ethically motivated steps to enhance mutually beneficial human-environment relations whose objective is to meet human needs while factoring nature. This includes costing environmental harm, accounting for the "production, use, and disposal of goods," upcycling waste, and enhancing biodiversity in pursuit of "a new kind of science that challenges conventional wisdom about power and priority" and whose ultimate aim is to fulfill basic human needs in cooperation with the natural world (Bjornerud 1997, 101). This latter example expands Longino's basic framework by incorporating the thesis that for human needs to be met, an ethical relation to the natural world is also necessary. When considered in conjunction with feminist contextual empiricism more generally, as well as with the other virtues, it fits nicely in Longino's larger conceptual ethos.

Now, if we consider these two examples as indicative of science that is oriented to meeting human needs, it quickly becomes clear that SO_2 geoengineering is unlikely to meet its requirements on the basis of scientific methodology, principles, practices, and objectives. Both caring science and the Gaia metaphor aim to meet human needs in tangible, practical ways with an understanding of human needs that is broad, intuitive, emotional, and not rigidly humanistic. On the other hand, climate engineering science, because its aims are also pragmatic, tends not to engage deeply with ethico-political issues aside from acknowledging their existence. For instance, in a widely cited article on stratospheric aerosols by Aswathy et al., the authors confess that "additional social and political conflicts between regions of the world might occur if it should come to discussions about the eventual implementation of SRM" (Aswathy et al. 2015 9608), but do not engage in any follow up or further analysis (Rasch et al. 2008). Most other geoengineering articles make no mention of politics, ethics, or policy at all (Jones et al. 2011; Niemeierr et al. 2011; Kalidindi et al. 2015).

Human Capabilities and Geoengineering

Yet, if we look at closely at geoengineering research, and particularly at the discourse used to justify its potential use, a case can be made that it, and those that support it, are interested in meeting human needs in ways that are non-trivial and persuasive. First, it is important to recognize that the stated primary objective of SO_2 geoengineering (and geoengineering in general) is to mitigate harmful climate disruption caused by human activity. Understood in this way, climate engineering clearly does aspire to meet the basic human need to live in a safe environment – particularly one in which temperature fluctuations and extreme climate events are regulated. Ostensibly, climate engineering also aims to ensure that future generations do not have to suffer with the deleterious effects of our past and present patterns of consumption and growth.

Yet interestingly, as stated in previous Chapters, those who argue in support of geoengineering tend to offer lukewarm rather than emphatic backing. Most proponents take the position that given the lack of political will and technical capacity for mitigation, coupled with such high levels of GHG concentrations and uncertainties associated with existing frameworks, research into geoengineering should continue with the assumption that it should only be deployed if absolutely necessary (Caldeira and Keith 2010; Blackstock et al. 2009; Klepper 2012). Even the Oxford Geoengineering Programme is careful in its call for further research stating that,

> Given the uncertainties and stakes involved, it is important that we conduct research to determine if any of the proposed geoengineering techniques could be employed without creating countervailing side-effects. It is far from certain that any of them could be, so it would be extremely unwise to rely on geoengineering as a 'silver bullet' for climate change. (Oxford Geoengineering Programme 2017).

In this tepid call for more research, however, discussion of the need to do so on the grounds of basic human interests remains widespread. The discourse of a tipping point is one such example wherein the argument is made that if we reach a climate tipping point, characterised by catastrophic climate events including a loss of Arctic sea ice, flooding, and extreme weather events, humanity's very existence could be imperilled. Geoengineering is accordingly proposed as a last resort response (Jackson 2009; Lenton and Vaughan 2009). As Lenton explain, the argument goes something like this,

> For many tipping elements, warning is unlikely to be early enough to allow aversive action by mitigation of long-lived greenhouse gases, notably carbon dioxide. It is conceivable that faster climate intervention methods, such as mitigation of short-lived radiative forcing agents or geoengineering to reduce incoming sunlight, could be more effective (Lenton 2011, 208).

There is an implicit appeal here to a generalized conception of human needs and wellbeing which, while failing to argue that that geoengineering is a social good in and of itself, does make the case that, overall, its use may be able to preserve conditions for a habitable world.

The contention that geoengineering constitutes a 'lesser evil' is closely associated with the tipping point claim used by cautious supporters of its use to meet human needs and potentially alleviate human suffering. This argument holds that because we are failing to adequately address climate change,

> …we may end up facing a choice between allowing catastrophic impacts to occur, or engaging in geoengineering. Both, it is conceded, are bad options. But engaging in geoengineering is less bad than allowing catastrophic climate change. Therefore, the argument continues, if we end up facing the choice, we should choose geoengineering (Gardiner 2009, 3).

This claim succinctly typifies the discursive balance between critique and support deployed by proponents of further climate engineering research. Its reasoning, to be clear, relies on an implicit appeal to human needs and interests by asserting that, in order to maintain a liveable world, it may by necessary to use geoengineering if we are faced with precisely this set of undesirable options.

The inclusion of arguments around cost-benefit analysis is another way in which human needs and wellbeing are obliquely addressed, and potentially served, by climate engineering. Research conducted using DICE, a standard cost-benefit model, has been used to make the case that SRM is indeed a plausible option which could result in benefits for humanity at large (Ban-Weiss and Caldeira 2010; Bickel and Agrawal 2012; Gramstad and Tjøtta 2010). The weighing of hazards and risks to welfare, relate to temperature change and associated climate events, form the substance of studies that support SRM climate engineering in this way. As Bickel argues, attempts to "rule CE out of bounds on ethical grounds, or to set impossible terms for its use, may do great harm since it could lead to catastrophic consequences harmful to our well-being" (Lane 2013, 156).

However, there is also a significant amount of researches that employ cost-benefit analysis to argue against geoengineering based on an appeal to harms that would undermine human flourishing and capabilities. In an extensive study of aerosol geoengineering using this approach, Goes et al. concludes that the risk of geoengineering leading to abrupt climate change if terminated is troublingly high, that cuts in CO_2 emissions fare better in any test, that climate engineering risks undermining the agency of future generations, and that "whether geoengineering is deployed in an economically optimal portfolio hinges on currently deeply uncertain assumptions" (Goes et al. 2011, 743). In an equally compelling piece by Alan Robock et al., the authors state that while,

Anthropogenic stratospheric aerosol injection would cool the planet, stop the melting of sea ice and land-based glaciers, slow sea level rise, and increase the terrestrial carbon sink, but produce regional drought, ozone depletion, less sunlight for solar power, and make skies less blue. Furthermore it would hamper Earth-based optical astronomy, do nothing to stop ocean acidification, and present many ethical and moral issues (Robock et al. 2009).

Interestingly, the Robock et al. piece represents a significant departure from most scientifically directed articles that tend not to engage concretely with issues related to human needs or interests. On the other hand, it is published in the *Geophysical Research Letters*, which aims to communicate scientific advances in an accessible way and thus has a wider remit than a scientifically mandated periodical. As an aside, journals like this could serve as a bridge between scientists and the public in relation to collaborative research and the deployment of feminist virtues in science.

It is noteworthy that, scientifically grounded arguments endorsing the need for climate engineering often assert its capacity to satisfy basic human interests only as a secondary issue. In contrast, the thematization of human needs has been much more explicit in media coverage of geoengineering. Many scientists have taken it upon themselves to engage with media on the subjects of ethics, policy and human interests as it relates to the politics of climate engineering. Social scientists have also begun to fill the void with compelling social, ethical, and politically engaged analyses in academic literature and the media. In arguing for more climate engineering research, scientists like David Keith, Ken Caldeira, and Ben Kravitz have been cited in publication like *Nature, The New York Times, The Guardian* and others where a human needs argument is made claiming that future research, and perhaps deployment, may be necessary given the severity of the climate challenge (Fountain 2017; Roston 2017; Mathieson 2015).

In a widely cited piece for *The New York Times* titled, "Panel Urges Research on Geoengineering," journalist Henry Fountain summarizes the findings of several prominent scientists at a National Academy of Sciences convened panel who jointly make the case for more research into geoengineering. Scientists like Ken Caldeira, Marcia K. McNutt, the editor of *Science*, Waleed Abdalati, a professor at the University of Colorado, as well as Raymond Pierrehumbert, a geophysicist at the University of Chicago, agreed that "It may be prudent to examine additional options for limiting the risks from climate change" since, at this point in time, it has "reached a point where the severity of the potential risks from climate change appears to outweigh the potential risks from the moral hazard" (Fountain 2015). The ability for geoengineering proponents to directly and accessibly make the case for further research in the name of serving the public good, even when journalists are careful to include critiques – as Fountain does –, marks an important means by which the human needs argument can be made. A similar case is made David Keith in an interview for the CBC (Canadian Broadcast Corporation) and an opinion piece for *Wired* in which he admits that geoengineering, while a "brutally ugly technical fix," (Wodskou 2014), may be ethically necessary. He asserts that there is "a clear moral case for researching geoengineering" given the "risk-risk tradeoffs we are presented with" (Keith and Wagner 2016).

Parallels can be drawn between public policy making around risky technologies in which an acceptable level of regulation has ensured that the development of particular scientific ideas and technological artefacts do not exceed socially acceptable and legal norms. Policy making itself has become more participatory, particularly with respect the environment, wherein human values that go beyond economics are given larger airing (Fisher et al. 2010). Participatory policy-making and public involvement in science has led to models of innovation and rule making that are inclusive and reflective of human concerns (Schot and Rip 1997).

A prime example of public policy innovation that has been relatively successful in ensuring the development of risky technologies are accompanied by sufficient levels of regulation is the development of genetic engineering. The history of public policy in this area suggest that, in relation to the application of these technologies to human populations, sound regulatory regimes can be adopted to ensure precaution is the norm but that scientific advance is able to continue (Carroll and Charo 2015; Hayakawa et al. 2016).

Current debates around the regulation of germline genetic editing technologies, which allows for the editing of DNA that is passed on to ones' offspring, is only the most recent of several innovations from IVF in the 80s to animal cloning, has had to be regulated in order to allow for foreseen and unforeseen negative consequences to be limited. Despite varying regulations between individual countries and the increasingly outsized role played by entrepreneurs and corporations in this area, there have been no catastrophic breaches of public trust even though disputes about pace, funding, access, and control continue to be of concern (Ledford 2015; Stock and Campbell 2000).

Whether this kind of participatory and socially responsible policy making is possible vis-à-vis geoengineering is uncertain. Currently, we are at the socio-technical stage at which decisions about its very value are being seriously discussed. Skepticism about the difficulty of instituting a regulatory regime in which all countries will sign on and uphold, as well as concerns regarding rogue experimentation and unintentionally providing an opening to other types of planetary interventions, makes regulation a secondary concern to questions about whether or not this is something we should do in the first place.

That being said, there are a number of frameworks being developed, including the Oxford Principles and those articulated by the Royal Society, which codify some of the most important norms around technological development including, for the Oxford Principles, public participation, the treatment of climate engineering as a public good, full transparency and public disclosure, independent assessment, and governance before deployment (Rayner et al. 2013; Gardiner 2011). These principles, while respectable as first steps, are quite abstract and do not adequately consider power, gender, race, class and a whole host of other factors.

One final and related means by which the argument can be made that geoengineering might serve the public good, and thus human interests, is by discursively constructing and materially regulating it as a public good. This novel argument is described in the UK House of Commons Select Committee on Science and Technology report in relation to arguments against patenting these technologies, which the authors contend would ensure that climate engineering is subject to strong public oversight such that private interests cannot "control whether and how geoengineering technology will be researched and used" (UK House of Commons 2010). Interpreted in this way, if geoengineering were constructed as a public technology, it might ensure that its capacity to fulfill human needs are not undermined by potential risks, dangers, and conflicts of interest. As Parthasarathy et al. puts it,

> Given the incredibly high stakes of geoengineering and climate change, a body with responsibility to the public—and with a broad, interdisciplinary, perspective—must use the power of the patent system to ensure the public good (Parthasarathy et al. 2010, 14).

Ostensibly, this would fulfill human need to live in an environment not subject to the health, human, and existential risks associated with climate change.

Yet, even a generous interpretation of this latter argument does not, either qualitatively or ethico-politically, adhere the interpretation of human needs feminist empiricist science calls for. Longino's final virtue is committed to an interpretation of human flourishing and scientific practice that "meets human needs and social needs; alleviates pain, suffering, and deprivation; and leads to the improvement of the material conditions of human life (Nagl 2005, 167). This goes much further than the preceding justification makes room for.

On the other side, a robust case can and has been made using ethico-political claims against climate engineering with sizeable attention paid to the subject of humanity's overall well-being (Gardiner 2011; Preston 2013; Jamieson 2009). The arguments in support of this position range from direct unintended consequences, the lack of governance structures, the top-down nature of the technology, the lack of transparency and public participation, as well as issues around intergeneration justice and the material requirements needed for authentic, qualitative change. Having discussed the majority of these at length, the latter two positions add an original perspective to the assertion that climate engineering does not, in fact, serve or address human needs and interests.

The subject of intergenerational justice or equity is a particularly fraught subject in the context of geoengineering since the decision to employ it clearly impinges on the agency and choice matrices of future generations. This is particularly the case with respect to SO_2 geoengineering for reasons that will be made clear after a brief discussion of the role of intergeneration justice in environmental discourse and law.

Historically, and in environmental law, the notion that the earth's resources must be safeguarded for future generations formally came into being in the early 1970s with the UNEP's Stockholm Declaration, which proclaims that,

> To defend and improve the human environment for present and future generations has become an imperative goal for mankind-a goal to be pursued together with, and in harmony with, the established and fundamental goals of peace and of worldwide economic and social development (UNEP 1972).

Subsequent agreements, including the UNFCCC, have also recognized these rights by highlighting the temporal character of environmental actions. In addition to considering the needs of future generations, intergenerational equity also must retain a 'conservation of options,' which means the current generation must not take actions that aim to predict what future generations might need or want (since this would impinge on their agency and violate a basic right to freedom) (Weiss 1990, 8). The predicate of avoiding economic and social activities that might irreparably damage the natural world and refraining from making decisions now that might undercut the freedoms of subsequent generations would be undermined by geoengineering.

In the case of sulphate geoengineering in particular, this kind of intergenerational justice stands a high likelihood of being violated and, as such, ultimately weakens the capacity for human needs and interests to be equitably realized. As Burns argues, the unintended consequences of SO_2 geoengineering, which could include ozone depletion and disruptions in precipitation levels, could physically damage the environment for future generations and thus violate principles of fairness and equity (Burns 2011). It might also undermine our current commitment to decarbonisation by lowering the perceived need to engage in mitigation efforts (this is often referred to as the 'moral hazard' argument), as well as leaving problems associated with termination effects in place. Specifically, SO_2 climate engineering poses a particular problem to human and intergeneration justice since, as models demonstrate, the risk of abruptly terminating its use would likely lead to a "buildup of carbon dioxide that had accrued in the atmosphere in the interim, with its suppressed warming effect, as well as the temporary suppression of climate-carbon feedbacks" (Burns 2011, 45). Burns cites a recent study which asserts that,

> [S]hould the engineered system later fail for technical or policy reasons, the downside is dramatic… The climate suppression has only been temporary, and the now CO_2-loaded atmosphere quickly bites back, leading to severe and rapid climate change with rates up to 20 times the current rate of warming of ≈0.2 °C per decade…(Brewer 2007 9951).

As such, decision-making capacity and the ability to exercise agency around whether or not to geoengineer by future generations is limited in light of inevitable technological lock-in. If we understand human needs and interests as inextricably linked to capacities like these, without which material needs also cannot be met, climate engineering cannot be adequately said to meet human interests in any meaningful way.

Svoboda et al. has taken this argument a step further in asserting that sulphate climate engineering also fails to meet the requirements of distributive and procedural justice since the fact that stable access to food and water will likely be disrupted for some violates the former (distributive justice), while the improbability that it will be deployed fairly and inexpensively jeopardizes the latter (procedural justice) (Svoboda et al. 2011). Significantly, it is also the case that the unintended side effects SO_2 climate engineering poses a direct challenge to both categories of justice.

Turning back to the present, what is often left out of geoengineering research and discourse is the fact that human needs, capacities, and interests are best realized when grounded, local experience is thematized. Scott (1998) pointed out the tendency for actions taken by international organizations, both public and private, to adopt a global gaze that conceals the human scale and deemphasizes "place, identity and community that actually constitute and constrain how humans interact with their environment" (Adger et al. 2010, 548). As maintained in previous Chapters, this is unquestionably something geoengineering research is guilty of reproducing. The consequence of this global gaze is that the needs and interests of local communities and groups including women, minorities, the poor, and the disenfranchised are overlooked or ignored entirely. Any techno-scientific endeavour that fails to adequately reflect local demands, desires and consequences cannot be said to adequately express or address human needs and interests – even when the argument is made that its' objective is to ensure a liveable world.

With respect to decision-making about alternative development models and their implementation, when ethics, policy making, and divergent priorities collide on local, national, and global levels, it becomes even more important to ensure participation in decision making through established newly articulated models is made possible. This might include free, prior, and informed consent, grassroots policy-making, citizen science, and engaged citizen participation (Ward 2011; Irwin 2002;

Beierle 2010). Forming venues through which to do this fits with the feminist approach taken here as well as democratic values more generally.

In a similar vein, David Demeritt points out that universalizing, abstract, and instrumental approaches to climate science, to which climate geoengineering subscribes, tends of mask its very value-ladenness which Longino's entire framework aims to make manifest. He argues that this "global scaling and universalizing appeal conceal the uneven political economy of GHG emissions by divorcing the problem of their accumulation in the atmosphere from related social and economic matters" (Demeritt 2001, 313). This critique of globalizing science can also be said to reflect an epistemological argument against climate engineering's ability to assist in human flourishing that fits with the assertion that its failure to do so is a result of a world view that conflicts with the belief that solutions to environmental problems are not to be "found in powerful technologies, such as geoengineering, but in more fundamental moral and political changes" (Scott 2012).

Theoretically, socially responsible science could be one way in which to inject geoengineering research with an orientation towards fulfilling human needs and capacities in the same way that it has been used to guide the objectives of genetic engineering (Sankar and Cho 2015; Bovenkerk 2015). Socially responsible scientific practices have become all the more urgent in light of the central insight that science is now postnormal – meaning that scientists themselves have to deal with unprecedented levels of scientific uncertainty and conflicting values coupled with the need for rapid decision making (Funtowicz and Ravetz 1992). This sense of urgency and uncertainty applies to geoengineering as well. Other disciplines have drawn on the discourse of SRS, as it has come to be known, in order to construct codes of ethics to guide controversial scientific practice. One example is the Uppsala Code of Ethics for Scientists, which articulates the following four principles set out by a group of prominent scientists at Uppsala University that aim to guide scientific research. Subsequent guidelines have build on these tenets (Jones 2007; Anderson 2007):

1. Research shall be so directed that its applications and other consequences do not cause significant ecological damage;
2. Research shall be so directed that its consequences do not render it more difficult for present and future generations to lead a secure existence. Scientific efforts shall therefore not aim at applications or skills for use in war or oppression. Nor shall research be so directed that its consequences conflict with basic human rights as expressed in international agreements on civic, political, economic, social and cultural rights;
3. The scientist has a special responsibility to assess carefully the consequences of his/her research, and to make them public; and
4. Scientists who form the judgement that the research which they are conducting or participating in is in conflict with this code, shall discontinue such research, and publicly state the reasons for their judgement. Such judgements shall take into consideration both the probability and the gravity of the negative consequences involved (Gustafsson et al. 1984, 312).

These principles, when combined with a capabilities approach as well as the basic definition of human needs and interests set out by Longino, forms a robust and feminist basis from which to make the claim that because of likely ecological effects, intergeneration harms, lack of representation, power differentials in decision-making, and private interests, climate engineering does not deliver on basic human needs. It is important to emphasize that feminist approaches to responsible science, aim to explicitly highlight those values and interests that have been a "chosen so as to meet the needs of society…[which]would be the morally justified political conditions under which scientific research would be pursued" (Kourany 2010, 68). There is, in fact, a proposed set of principles of ethics for geoengineering research generated by researchers from IASS Potsdam in association with the Institute for Science, Innovation and Society at the University of Oxford, which articulates an extensive set of articles aimed at governance based on existing principles of international environmental law. Also addressed are guidelines around scientific conduct, environmental assessment, scientific attributes, research, public participation and authorization (IASS Working Paper 2015). While it is not possible to thoroughly assess the document here, it is important to point out that the objective of the code is to proactively create the conditions under which geoengineering might become more socially responsive.

Another critical way in which to reflect on human needs and interests in light of geoengineering can be found a review penned by Longino of Philip Kitcher's work. In her review, she draws attention to a conflict related to human needs that is especially relevant to geoengineering. She points out that private, corporate actors are increasingly conducting scientific experimentation with consequences that undermine global and local desires and interests. This tendency is directly applicable to climate engineering where the private interests of individuals (Russ George, Bill Gates), corporations (Royal Dutch Shell, and Exxon), as well as think tanks and industry-funded groups (AEI and the Heartland Institute – which is partially funded by ExxonMobil), are guiding research priorities. I return to this further on.

Because, it is the case that geoengineering of all sorts, but sulphate SRM in particular, holds a globalizing ethos that conflicts with the feminist principles of local control and marginalized interests, Longino maintains that human needs will continue to be unmet by traditional science. This is especially the case when questions like "How can individual societies or communities maintain the control envisaged? And what happens when different equally advantaged or even differently advantaged societies embrace values and agendas that will conflict when put into action?" (Longino 2002a, b, c, 568), are not a part of the agenda. Longino also contends that, this virtue requires research into areas traditionally neglected, thereby "triggering the novelty criterion in its weaker interpretation" (Longino, 389). What I take her to mean by this is that the form of novelty this particular virtue cultivates does not necessarily require completely new theories, but a generalized move towards scientific practices that pay attention to what has been traditionally ignored. According to this interpretation, it would include neglected populations, entities, phenomena, theories, principles and methodologies. As discussed in previous Chapters, this is an essential area in which climate engineering research, particularly at the level of science, is lacking. The categories of gender, race, class, geography, power, geography, and temporality represent significant sluices of analysis that are both novel and representative of those groups for whom human interests are not likely to be met. While research aimed at contextualizing geoengineering at the levels of policy, political economy and discourse have begun to introduce novel views and perspectives, it is this focus on the disempowered, disenfranchised, overlooked and negatively impacted, whether materially, politically or socially, that requires additional work.

One of the central guiding principles of scientific practice that Longino is patently clear about is that scientists must be precluded from profiting commercially from their work. She states that the "epistemological justification for this" is that "scientists ought not to have a stake in the outcome of their research since, scientists being human, such a stake might bias their interpretations..."(Longino 1990, 88). As such, protocols that ensure rigorous transparency, like making the science publicly accessible and the sources of funding transparent, is essential. This is particularly true in the case of climate engineering where distrust of process and outcome is high (Winickoff and Brown 2013).

This kind of transparency, which must be carried out in order to satisfy human interests, has generally not been followed vis-à-vis geoengineering. However, it is important to emphasize that transparency in the context of wicked postnormal science is highly complicated and difficult to achieve. It is also the case that there is little direct evidence that scientists themselves are profiting off of geoengineering research or technology. Rather, the problem seems to be the complicated and rather opaque funding mechanisms through which research is conducted and conferences sponsored, the presence of prominent scientists on the boards of entities interested in geoengineering, and the corporate sponsoring of the research itself.

Prominent examples of the latter two can be seen in the cases of Harvard scientist David Keith who is on the board of Carbon Engineering, a CO_2 capture investment firm to which Bill Gates has given money, and Ken Caldeira, another scientist who teaches at Stanford, and who has conducted research for Intellectual Ventures, a private research firm, and is listed as an 'inventor' on their website (Vidal 2012). A further example comes from work done by the chief scientist at BP, Steven Koonin, who, according to geoengineering critic Clive Hamilton, in 2009 led research that resulted in the publication of a report on climate engineering for Novim Group – a not-for-profit scientific research company. As Gardiner states,

> The authors felt the need to declare that in playing a prominent role Koonin had no conflict of interest, arguing, implausibly, that it is not possible to say that promoting research into geoengineering has any bearing on policies to reduce carbon dioxide emissions and thus BP's bottom line. In 2009 Koonin was appointed Under-Secretary for Science at the United States Department of Energy (Hamilton 2013).

Overall, the impact of this kind of opacity and perceived conflict of interests around geoengineering is that it acts as a further barrier to fulfilling human interests as well as the need for transparency in science.

Cumulatively, this Chapter has addressed a number of issues relating to the virtue of human needs with regard to geoengineering beginning with mechanisms through which scientific support by proponents have harnessed arguments in favor of further research using the tipping point, lesser evil and cost-benefit as discursive frames. A cursory look at media coverage shows more engagement with the idea climate engineering might contribute to the public good by making the world more habitable and prosperous.

Yet, as made clear, there is a strong case to be made that geoengineering does not fulfill human needs or extend human capacities with respect to ethical norms, intergenerational justice, globalization, and corporate conflicts of interest. It is important to keep in mind that, on a very basic level, applicability to human needs is a virtue that must be oriented to improving the human condition, ensuring feminist objectives are represented, and guaranteeing that gender is not disappeared. Understood in this way, climate engineering does not meet these criteria.

It is also critical to keep in mind that Longino's underlying argument is that the applicability to human needs virtue, in addition to all the other principles she articulates, are feminist and that while gender not be explicitly manifest, what is important is encouraging the practice of doing science as a feminist. As such, feminist politics and knowledge producing practices, which are constituted by the virtues of empirical adequacy, novelty, heterogeneity, diffusion of power, mutuality of interaction and human needs, is aimed at the production of an accountable, committed, *and feminist* science as everyday practice.

The next, and final, Chapter provides a synthesis of the FCE case made against geoengineering thus far. Also discussed out are some of the shortcomings of this approach, which I contend can be resolved using theoretical and methodological contributions from feminist standpoint theory and technofeminist approaches to technology.

References

Adger, W. N., et al. (2010). Progress in environmental change. *Global Environmental Change, 20*(4), 547–549.

Anderson, M. S. (2007). Collective openness and other recommendations for the promotion of research integrity. *Science and Engineering Ethics, 13*(4), 387–394.

Aswathy, V. N., et al. (2015). Climate extremes in multi-model simulations of stratospheric aerosol and marine cloud brightening climate engineering. Atmospheric Chemistry and Physics. *European Geosciences Union, 15*(16), 9593–9610.

Ban-Weiss, G. A., & Caldeira, K. (2010). Geoengineering as an optimization problem. *Environmental Research Letters, 5*(3). http://iopscience.iop.org/article/10.1088/1748-9326/5/3/034009/meta. Accessed 23 Feb 2017

Beierle, T. C. (2010). *Democracy in practice: Public participation in environmental decisions.* London: Routledge.

Bickel, J., & Agrawal, S. (2012). Reexamining the economics of aerosol geoengineering. *Climatic Change, 119*, 993–1006.

Bjornerud, M. (1997). Gaia: Gender and scientific representations of the Earth. *NWSA Journal, 9*(3), 89–106.

Blackstock, J. J., et al. (2009). *Climate engineering responses to climate emergencies*. arXiv preprint arXiv:0907.5140. https://arxiv.org/pdf/0907.5140.pdf. Accessed 22 Feb 2017.

Bovenkerk, B. (2015). Scientific responsibility: Should analysis start with the scientists? *The American Journal of Bioethics, 15*(12), 66–68.

Brewer, P. G. (2007). Evaluating a technological fix for climate. *Proceedings of the National Academy of Sciences, 104*(24), 9915–9916.

Burns, W. C. (2011). Climate geoengineering: Solar radiation management and its implications for intergenerational equity. *Stanford Journal for Law and Policy*.http://web.stanford.edu/group/sjlsp/cgi-bin/orange_web/users_images/pdfs/61_Burns%20Final.pdf. Accessed 24 Feb 2017.

Caldeira, K., & Keith, D. W. (2010). The need for climate engineering research. *Issues in Science and Technology, 27*(1), 57–62.

Carroll, D., & Charo, R. A. (2015). The societal opportunities and challenges of genome editing. *Genome Biology, 16*(1), 242.

Demeritt, D. (2001). The construction of global warming and the politics of science. *Annals of the association of American geographers, 91*(2), 307–337.

Eriksson, K. (2002). Caring science in a new key. *Nursing Science Quarterly, 15*(1), 61–65.

Fisher, E., et al. (2010). The public value of nanotechnology? *Scientometrics, 85*(1), 29–39.

Fountain, H. (2015, February 10). Panel urges research on geoengineering as a tool against climate change. *The New York Times*. http://www.nytimes.com/2015/02/11/science/panel-urges-more-research-on-geoengineering-as-a-tool-against-climate-change.html. Accessed 15 Sept 2016.

Fountain, H. (2017). White House urges research on geoengineering to combat global warming. *New York Times*, 10 January 2017. https://www.nytimes.com/2017/01/10/science/geoengineering-climate-change-global-warming.html?_r=0. Accessed 21 Feb 2017.

Funtowicz, S. O., & Ravetz, J. R. (1992). Three types of risk assessment and the emergence of post-normal science. In S. Krimsky & D. Golding (Eds.), *Social Theories of Risk* (pp. 251–273). Westport: Praeger.

Gardiner, S. (2009). *Is arming the future' with geoengineering really the lesser evil?* Some doubts about the ethics of intentionally manipulating the climate system. http://folk.uio.no/gasheim/Gar2010b.pdf. Accessed 15 Feb 2017.

Gardiner, S. M. (2011, May 1). Some early ethics of geoengineering the climate: A commentary on the values of the Royal Society report. *Environmental Values, 20*(2), 163–88.

Goes, M., Tuana, N., & Keller, K. (2011). The economics (or lack thereof) of aerosol geoengineering. *Climatic Change, 109*(3–4), 719–744.

Gramstad, K., & Tjøtta, S. (2010). *Climate engineering: Cost benefit and beyond*. https://mpra.ub.uni-muenchen.de/27302/. Accessed 23 Feb 2017.

Gustafsson, et al. (1984). The Uppsaa code of ethics for scientists. *Journal of pease Research, 21*(4), 311–316.

Hamilton, C. (2013). *How Bill Gates is engineering the Earth to resist climate change*. Crikey, 26 Feb 2013. https://www.crikey.com.au/2013/02/26/how-bill-gates-is-engineering-the-earth-to-resist-climate-change/. Accessed 23 Feb 2017.

Hayakawa, T., et al. (2016). Report of the international regulatory forum on human cell therapy and gene therapy products. *Biologicals, 44*(5), 467–479.

IASS Working Paper. (2015). *Responsible scientific research involving geoengineering*. Institute for Advanced Sustainability Studies (IASS), Postdam. http://www.insis.ox.ac.uk/fileadmin/images/misc/An_Exploration_of_a_Code_of_Conduct.pdf. Accessed 25 Feb 2017.

Irwin, A. (2002). *Citizen science: A study of people, expertise and sustainable development*. London: Routledge.

Jackson, S. C. (2009). Parallel pursuit of near-term and long-term climate mitigation. *Science, 326*, 526–527.

Jamieson, D. (2009). The ethics of geoengineering. *People and Place, 1*, 2. https://philpapers.org/rec/JAMTEO-8. Accessed 24 Feb 2017.

Jones, N. L. (2007). A code of ethics for the life sciences. *Science and engineering ethics, 13*(1), 25–43.

Jones, A., et al. (2011). comparison of the climate impacts of geoengineering by stratospheric SO_2 injection and by brightening of marine stratocumulus cloud. *Atmospheric Science Letters, 12*(2), 176–183.

Kalidindi, S., et al. (2015). Modeling of solar radiation management: A comparison of simulations using reduced solar constant and stratospheric sulphate aerosols. *Climate Dynamics, 44*(9-10), 2909–2925.

Keith, D., & Wagner, G. (2016). To help cool the climate, add aerosol. *Wired Magazine*, 10 May 2016. https://www.wired.com/2016/10/help-cool-climate-add-aerosol/. Accessed 1 Mar 2017.

Klepper, G. (2012). What are the costs and benefits of climate engineering? And can we assess them? *Sicherheit und Frieden (S+ F)/Security and Peace, 30*, 211–214.

Kourany, J. A. (2010). *Philosophy of science after feminism.* Oxford: Oxford Univerity Press.

Lane, L. (2013). Climate engineering and the Anthropocene era. In W. C. G. Burns & A. L. Strauss (Eds.), *Climate change geoengineering: Philosophical perspectives, legal issues, and governance frameworks* (pp. 115–145). New York: Cambridge University Press.

Ledford, H. (2015). Where in the world could the first CRISPR baby be born? *Nature.* Available at: http://www.nature.com/news/where-in-the-world-could-the-first-crispr-baby-be-born-1.18542. Accessed 4 May 2018.

Lenton, T. M. (2011). Early warning of climate tipping points. *Nature Climate Change, 1*(4), 201–209.

Lenton, T. M., & Vaughan, N. E. (2009). The radiative forcing potential of different climate geoengineering options. *Atmospheric Chemistry and Physics, 9*(15), 5539–5561.

Lindholm, L., & Eriksson, K. (1993). To understand and to alleviate suffering in a caring culture. *Journal of Advanced Nursing, 18*, 1351–1361.

Longino, H. E. (1990). *Science as social knowledge: Values and objectivity in scientific inquiry.* Princeton: Princeton University Press.

Longino, H. E. (1996). Cognitive and non-cognitive values in science: Rethinking the dichotomy. In L. H. Nelson & J. Nelson (Eds.), *Feminism, science, and the philosophy of science* (pp. 39–58). Dordrecht: Kluwer Academic.

Longino, H. (2002a). Reply to Philip Kitcher. *Philosophy of Science, 69*(4), 573–577.

Longino, H. (2002b). Science and the common good: Thoughts on Philip Kitcher's science, truth, and democracy. *Philosophy of Science, 69*(4), 560–568.

Longino, H. E. (2002c). *The fate of knowledge.* Princeton/Oxford: Princeton University Press.

Longino, H. E., & Lennon, K. (1997). Feminist epistemology as a local epistemology. *Proceedings of the Aristotelian Society*, Supplementary Volumes, *71*, 19–54.

Mathieson, K. (2015). Is geoengineering a bad idea? *The Guardian*, 11 February 2015. https://www.theguardian.com/environment/2015/feb/11/is-geoengineering-a-bad-idea-climate-change. Accessed 21 Feb 2017.

Nagl, S. (2005). Biomedicine and moral agency in a complex world. In M. Shildrick & R. Mykitiuk (Eds.), *Ethics of the body: Postconventional chanllenges* (pp. 155–174). London: MIT Press.

Niemeier, U. H., et al. (2011). The dependency of geoengineered sulfate aerosol on the emission strategy. *Atmospheric Science Letters, 12*(2), 189–194.

Nussbaum, M. (1999). Women and equality: The capabilities approach. In M. F. Loutfi (Ed.), *Women, gender and work: What is equality and how do we get there?* (pp. 45–68). Geneva: International Labour Office.

Oxford Geoengineering Programme. (2017). *Why consider it?* Oxford Geoengineering Programme. http://www.geoengineering.ox.ac.uk/what-is-geoengineering/why-consider-geoengineering/. Accessed 19 Feb 2017.

Parthasarathy, S., et al. (2010). *A public good?* Geoengineering and intellectual property. Gerald R. Ford School of Public Policy Working Paper, Michigan. http://www.stpp.fordschool.umich.edu/policy-consultations/GAO%20papers/Item%20B15-A%20Public%20Good%20GAO%20STPP%20Working%20Paper%2010-1.pdf. Accessed 24 Feb 2017.

Preston, C. J. (2013). Ethics and geoengineering: Reviewing the moral issues raised by solar radiation management and carbon dioxide removal. *Wiley Interdisciplinary Reviews: Climate Change, 4*(1), 23–37.

Rasch, P. J, Tilmes, S, Turco, R. P, Robock, A Oman, L, Chen, C, Stenchikov, G. L, & Garcia, R. R. (2008, November). An overview of geoengineering of climate using stratospheric sulphate aerosols. *Philosophical Transactions. Series A Mathematical, Physical, and Engineering Sciences, 366*(1882), 4007–4037.

Rayner, S., et al. (2013). The Oxford principles. *Climatic Change, 121*(3), 499–512.

Robock, A., et al. (2009). Benefits, risks, and costs of stratospheric geoengineering. *Geophysical Research Letters 36*, 19. http://onlinelibrary.wiley.com/doi/10.1029/2009GL039209/full. Accessed 23 Feb 2017.

Roston, E. (2017). *Scientists want to give the atmosphere an antacid to relieve climate change.* Bloomberg 12 December 2016. https://www.bloomberg.com/news/articles/2016-12-12/scientists-want-to-give-the-atmosphere-an-antacid-to-relieve-climate-change. Accessed 21 Feb 2017.

Sankar, P. L., & Cho, M. K. (2015). Engineering values into genetic engineering: A proposed analytic framework for scientific social responsibility. *The American Journal of Bioethics, 15*(12), 18–24.

Schot, J., & Rip, A. (1997). The past and future of constructive technology assessment. *Technological Forecasting & Social Change, 54*, 251–268.

Scott, J. C. (1998). *Seeing like a state: How certain schemes to improve the human condition have failed.* New Haven: Yale University Press.

Scott, D. (2012). Geoengineering and environmental ethics. *Nature Education Knowledge, 3*, 10. http://www.nature.com/scitable/knowledge/library/geoengineering-andenvironmental-ethics-80061230. Accessed 12 Feb 2017.

Stock, G., & Campbell, J. (2000). *Engineering the human germline.* New York: Oxford University Press.

Svoboda, T., et al. (2011). Sulfate aerosol geoengineering: The question of justice. *Public Affairs Quarterly, 25*(3), 157–179.

UK House of Commons: Select Committee on Science and Technology. (2010). *The regulation of geoengineering.* London: UK House of Commons Stationary Office.

UNEP. (1972). *Declaration of the united nations conference on the human environment.* United Nations Environmental Program. http://www.unep.org/documents.multilingual/default.asp?documentid=97&articleid=1503. Accessed 24 Feb 2017.

Vidal, J. (2012). Bill Gates backs climate scientists lobbying for large-scale geoengineering. *The Guardian*, 6 February 2012. https://www.theguardian.com/environment/2012/feb/06/bill-gates-climate-scientists-geoengineering. Accessed 13 Feb 2017.

Ward, T. (2011). The right to free, prior, and informed consent: Indigenous peoples' participation rights within international law. *Northwestern University Journal of International Human Rights, 10*, 54–84.

Watson, J. (2007). Watson's theory of human caring and subjective living experiences: Carative factors/caritas processes as a disciplinary guide to the professional nursing practice. *Texto & Contexto-Enfermagem, 16*(1), 129–135.

Watson, J. (2012). Viewpoint: Caring science meets heart science: A guide to authentic caring practice. *American Nurse Today, 7*, 8. https://www.americannursetoday.com/viewpoint-caring-science-meets-heart-science-a-guide-to-authentic-caring-practice/ August 2012. Accessed 19 Feb 2017.

Watson, J., & Smith, M. C. (2002). Caring science and the science of unitary human beings: A trans-theoretical discourse for nursing knowledge development. *Journal of Advanced Nursing, 37*(5), 452–461.

Weiss, E. B. (1990). In fairness to future generations. *Environment: Science and Policy for Sustainable Development, 32*(3), 6–31.

Winickoff, D. E., & Brown, M. B. (2013). Time for a government advisory committee on geoengineering research. *Issues in Science and Technology, 29*(4), 79–85.

Wodskou, C. (2014). Give geoengineering a chance to fix climate change: David Keith. *CBC News*, 29 March 2014. http://www.cbc.ca/news/technology/give-geoengineering-a-chance-to-fix-climate-change-david-keith-1.2586882. Accessed 1 Mar 2017.

Chapter 9
Conclusion

Abstract This final Chapter, summarizes some of the book's most significant findings and undertakes a discussion of feminist standpoint theory and technofeminism. It make the case that certain insights from each can act as a supplement to FCE, which tends to fall short on the critical subjects of gender and power, discourse, and design. Incorporating an account of how climate engineering can be studied from these perspective adds an element of interdisciplinarity to FCE which, in accordance with its virtues of novelty and heterogeneity, should be encouraged.

Keywords Helen Longino · Feminist standpoint theory · Technofeminism · Marginalization · Power · Privilege · Democratic

In this final Chapter, before summarizing some of the most significant findings and offering some brief closing remarks, I undertake a short discussion of feminist standpoint theory and technofeminism and make the case that certain insights from each can act as a supplement to FCE, which tends to fall short on the critical subjects of gender and power, discourse, and design. Incorporating an account of how climate engineering can be studied from these perspective adds an element of interdisciplinarity to FCE which, in accordance with its virtues of novelty and heterogeneity, would be encouraged.

Longino's framework, and feminist empiricism in general, has been criticized on a number of levels from ignoring the importance of evidential objectivity, elevating a 'feminine,' rather than a feminist, epistemology, and undermining and/or holding too close to the value of truth (Pinnick et al. 2003; Haack 1996, 2009; Buckler 2010; Hawkesworth 2006). The objective here is not to examine the complexity of these arguments but to explore two further streams of feminist science studies that have the potential to fill a few theoretical gaps that have left feminist empiricism open to criticism. In what follows, I detail the contributions made by Sandra Harding's feminist standpoint theory and Judy Wajcman's technofeminist framework to the study of science and technology particularly with respect to the ways in which they can strengthen the feminist critique of geoengineering.

Feminist Standpoint Theory

The critique of Longino made by scholars of science that subscribe to a feminist standpoint perspective stems from the question of whether her framework is, first, sufficiently feminist, and second, whether it retains too much positivism. These critiques are particularly applicable as it relates to the practice of science and the unique contributions women can make. As one of the most compelling frameworks that reflects this perspective, it is articulated best by feminist scholars like Sandra Harding (1986a, b, 1992, 2004a, b, 2015), Nancy Hartsock (1998, 2004), Susan Hekman (1999, 2013), and Donna Haraway (1988, 1994). Susan Harding in particular addresses the problem of latent positivism by arguing that the context of justification, which include the scientific method itself, needs to be *more* critically examined since it is often "the norms of inquiry" that "results in androcentric results" (Harding 1989, 26).

It is important to recognize that, although FST and FCE deal with similar subject matter, they are markedly different with respect to their epistemological assumptions, ontological views, and social categories. The purpose of including FST in this discussion is not to engage in a comparative analysis but specifically to examine what the concept of standpoint might offer in the form of a critique of the science that underlies geoengineering and our (the public's) position on its possible use. As such, it is with respect to the specific insights standpoint theory makes to understanding the relationship between gender and science, as well as the role marginalization plays in epistemology, that adds something novel to the evaluation of geoengineering left out by Longino. In what follows, a brief overview of Sandra Harding's approach to standpoint theory is given, followed by a succinct discussion of what it contributes that is different from the feminist analysis of climate engineering that has been made using FCE.

For Harding, it is not sufficient to assert that inclusive and heterogenous communities that make room for the traditionally marginalized viewpoints leads to better science, but that it is *the taking* of these perspectives that produces better science. Harding contends that marginalized groups, particularly but not limited to women of color, because of their experience of "different social and cultural locations, and in the context of local and global systems of empowerment, oppression, and exploitation," are uniquely positioned to "reveal the objective natures and conditions of dominant groups" (Harding 2008, 14). She calls this practice 'science from below' – an explicitly feminist research practice that elevates marginalized knowledge – and calls into question dominant understandings of health, biology, and the environment as well as the hierarchical, misogynistic, and oppressive structures they are a product of. The assertion that knowledge produced by those socially situated at the margins is qualitatively 'better,' i.e. less distorted and partial, than those at the center is standpoint theory's most controversial claim.

Very basic examples taken from environmental science include projects that give prominence to the knowledge of local fisherman regarding the preservation and management of fishstocks, and of women in indigenous communities who play critical roles in monitoring biodiversity (Moller et al. 2004; Kofinas 2002). In the case of fish, research has been done that calls for women to be incorporated into fishing management based on their experiences in the industry (Neis and Williams 1997). Similarly, with respect to climate change, a standpoint perspective would entail a radical rethinking of knowledge practices around our relation with the natural world by elevating the views, experiences and expertise of women in particular. While women tend to have less power in decision-making and resource ownership, their tacit knowledge of how to manage available resources in line with familial needs is invaluable (Neumann and Hirsch 2000; Brown 2011). However, while helpful, under close scrutiny, it would appear that, this interpretation of standpoint theory is not substantively dissimilar from Longino's FCE. While the elevation of women's knowledge and experiences as an end in itself is not an objective of feminist empiricism, it often is the case that it is an outcome of contextual virtues.

For standpoint theory to contribute something novel, it has to be interpreted in a much more radical way. This would require that inclusion extends beyond diversity to incorporate, in particular, those perspectives that "undermine hierarchical power structures and counteract the negative effects of oppression on knowledge production" (Intemann 2010, 791). Standpoint theory is ethically committed to dismantling power structures as a part of scientific practice. Understood in this way, climate science would need to be radically rethought – i.e. by challenging masculinized norms that include control over nature, that gage risk "through a prism of privilege, power, and geography" (Seager 2009, 16), and where market logic (cost/benefit; winners/loser) reigns. Central to this is a reframing of epistemology consistent with a discourse of climate in/justice where climate change itself is seen as both a cause and consequence of unequal power relations (Lykke 2009; Gaard 2011).

In the context of geoengineering science, also required is the inclusion and explicit incorporation/elevation of non-traditional knowledge formations into the study of climate change based on standpoint theory's contention that alternative and experience based epistemologies produce superior and more robust knowledge. For standpoint feminists, this includes paying attention "to everyday, routine, and often mundane activities that provide different opportunities for 'seeing' how social relations are shaped by power, and how responsibility and action are placed on differently and unequally situated bodies" (Bee et al. 2015, 5). Experiential and non-scientific knowledge that reject universalizing logics and embraces material and felt sensation are key.

To this, Bee et al. (2015) also includes feminist practices that undercut "masculine narratives of control and dominance," and highlights "the corporeal and embodied implications of climate change," as "something visceral, material, embodied and part of everyday" (Bee et al. 2015, 344). Climate engineering science, from this perspective, would be dismissed as inimical to standpoint feminists in light of its discounting of any of these practices as well as its emphasis is on globalizing narratives, capitalistic understandings of the economy (carbon markets, profitability, cost/benefit analysis), undertheorization of power relations, detachment from race, gender, class and socio-economic analyses, and tendency to elide the responsibility of the state and corporate actors.

On the subject of power in particular, the science underpinning climate engineering, from a standpoint perspective, would demand a clear theorization of how power relations might work to foster inequality in the context of its potential use. A well-defined and consistent sense of how geoengineering might exacerbate "injustices in material conditions and normative expressions, within societal structures and institutions of various kinds," which are then "lived, expressed, and reproduced through social practices" (Kaijser and Kronsell 2014) is essential. Understanding the various ways in which this kind of theorization could be made manifest remains necessary.

It is a robust theory of power and knowledge which Harding pushes to the fore and which Longino lacks that constitutes the most significant contribution of standpoint theory. It is unearthing practices of suppression, re-conceptualizing women as "active agents in in the process of scientific and technological change" (Tuana 2013, 25), and pushing forward intersectional research practices that constitute standpoint theory's central contribution.

Before turning to the insights of technofeminism, some clarification is required regarding the oft-cited critique of standpoint theory in terms of its capacity to produce objective and reliable knowledge out of such a seemingly subjective framework. Remember that feminist contextual empiricism retains a conception of objectivity based on the outcomes of social and intersubjective processes of debate and criticism (Longino 1990, 76). Longino's FCE asserts that background assumptions and constitutive values are not barriers against objectivity given that what makes a theory objective is that it emerges out of "practices of inquiry [that] are not individual but social" (Longino 1990, 67). Yet Longino also argues against the proposition that mere fact of marginalization makes for better science.

Conversely, standpoint theory begins from a different position in aiming to undercut the objectivity/subjectivity, realist/constructivist divides that pervade established science, and, in doing so, reconceptualize Western notions of value neutral objectivity as rooted in social inequality. Harding's notion of 'strong objectivity,' on the other hand, asserts that "all human thought can only be partial"(Harding 1995, 341), that knowledge is always situated and contextual, and that knowledge generated from outside traditional frameworks are 'objective' in the sense that they promotes growth in ways that are socially transformational and just (Harding 2015).

Thus, what standpoint theory offers to the study of climate engineering is a more pointed and critical perspective in which the central thesis is that it is not sufficient to point out that women and other marginalized groups must be incorporated into science, evaluation, policy making, and funding mechanisms, but that it is their knowledge, experiences, insights, and questions that deserve to take center stage. While FCE does address the effects of marginalization in knowledge formation, (Longino is clear that scientific communities must be a priori inclusive), it does not go that extra step in addressing how it might be that "precisely because…the mainstream community has excluded the dissenting views from consideration" that this "constitutes evidence of the absence of objectivity" (Levisohn 2001, 343). The inclusion and study of the effects of relations of power, critical consciousness, marginalization and privilege in the production of scientific knowledge would not only benefit FCE, but when applied to analyze climate science and geoengineering also yield beneficial, insightful, novel and heterogenous outcomes – even if it results in the rejection of geoengineering itself.

Technofeminism

The second feminist framework that can invigorate FCE, particularly with respect to the way it has been used to assess and critique geoengineering, is technofeminism. As stated, geoengineering is unique in that it exists in this liminal space as both speculative science and technological artefact. While FCE and FST address the science extensively, it is the technology aspect that often falls through the cracks. Technofeminism, also referred to as feminist technoscience, is an expansive and interdisciplinary field of research aimed at studying the role gender plays in media, information studies, and biosciences using resources from science and technology studies, feminist science studies, and postmodernism. As articulated by scholars like Judy Wajcman (2000, 2000), Anneke Smelick (2008), Cynthia Cockburn (1992), and Donna Haraway (1990, 2013), technofeminism works to find a middle ground between the celebration of the technology's emancipatory potential and its critique. This perspective treats technology as an artefact that is social, technical, and bound up in a mutually shaping relationship with gender. Gender relations, in this context, are reflected in technology on socio-technical and material levels such that "gendered identities and discourses as [seen as being] produced simultaneously with technologies" (Wajcman 2007, 293). Put another way, as Wajcman, argues, "Technology is then understood as both a source and consequence of gender relations" (Wajcman 2006, 15).

It is this theory of mutual shaping and co-construction that is useful to the study of climate engineering that both compliments and extends the insights proffered by feminist empiricism. What technofeminism does is to demonstrate how the process of gendering occurs at a material level in addition to social and discursive ones (Wajcman 2007, 294). It is on the level of discourse and culture that is of particular interest since FCE does not explicitly engage with discourse as field of analysis. A comprehensive accounting of the role discourse plays in justifying geoengineering is central to the way in which technologies develop and how they are ultimately used. A useful example Wajcman gives is with respect to the way in which sex differences in the medical sciences have been constructed through gendered discourse in medical manuals, teaching techniques, and via medical advice. Significantly, it was precisely this discourse that gave rise to and secured how modern medicine came to distinguish between male and female bodies (Wajcman 2002).

Regarding geoengineering, the insight that society and technology, and thus gender and technology, are mutually constitutive, that provokes the need for an in depth and critical analysis of the discourse surrounding its construction as a meaningful techno-semiotic set of practices and techniques. The way in which scientific studies of SO2 geoengineering frame it as ethically neutral, inherently progressive, and imperative are significant areas of future research (Ellul 1964; Postman 1993; Winner 1977). The further study of media representations, both visual and linguistic, are also noteworthy and could supplement FCE by demonstrating precisely how the values, norms, and assumptions that underpin the science of geoengineering are deployed socially and culturally.

Technofeminism also encourages a view of technology as always in the process of becoming – which is to say, not fixed. This opens space for the central insight that because technologies are social artefacts, they have the potential to be changed and transformed. As Wajcman argues, "Different groups of people involved with a technology can have different understandings of that technology, and consumers or users can radically alter the meaning and deployment of technologies" (Wajcman 2006, 15). Although the feminist reading of geoengineering thus far has been critical of its potential use and deployment by articulating all the reasons it does not reflect feminist scientific virtues and principles, technofeminism encourages thinking about the ways in which current and developing technologies might be shaped towards more just and prosocial ends.

While it is the case that contemporary technofeminism has been used to examine and explain the contingent and flexible nature of the biological sciences and ICTs (the Internet in particular), it is not inconceivable that geoengineering's basic principles could also be adapted and transformed by interested parties, including women, towards less invasive techniques. Geotherapy, for example (Goreau et al. 2014), takes some of the central insights from geoengineering to develop and extend processes already rooted in the earth's natural systems. This includes things like sustainable agriculture, producing polymers for use in construction material that store carbon, developing planting techniques for desert environments etc. This approach leaves the window open for the possibility that climate engineering techniques and practices could be transformed into a more heterogenous set of technologies capable of addressing human needs.

Technofeminist analysis also includes the examination of the ways in which technological design often fails to reflect the interests of women in any substantive way. Wajcman draws on Anne-Jorunn Berg's 1994 study of smart homes as reflecting gendered, specifically male, interests by "centralize[ing] control and regulation of all function in the local network or 'house-brain" (Wajcman 2002, 357), thereby ignoring the role of house and care work. In this case, the needs and interests of the mostly male designers were materially built into process of innovation – much like they are in geoengineering technologies that, without extending the analogy too far, have been designed with similar values and concerns in mind. Looking at the ways in which technological design is gendered could yield important insights by revealing how S02 climate engineering is more than just a technical artefact since its design has reflects "technical, economic, organizational, political, and even cultural elements" (Wajcman 2002, 352).

In examining climate engineering using feminist science studies, what has become patently clear is that there are barriers to using a gendered lens in order to study the physical and environmental sciences. This is the case not only because of the circuitous and implicit ways in which issues around gender express themselves, but also because of the way in which the practice of feminist science is traditionally seen as requiring an essentialized understanding of women. This is not the kind of analysis Longino is engaged in. Her feminist virtues of empirical adequacy, ontological heterogeneity, novelty, mutuality of interaction, and attending to human needs emerge out of a desire to account for how science has been traditionally conducted and, in establishing its grounding in values and interests, offer a framework through which to guide scientific practice in ways that thematize women's experiences that are in "pursuit of ends that include a heterogenous, equitable, novel and democratic science, a feminist science" (Bardzell and Churchill 2011).

Taken as a whole, the use of Longino's framework to serve as a foundation for this critique of climate engineering lies its unique ability to engage in building an approach to epistemology that incorporates values, interests, gender, and social practice into the production of knowledge without engaging in essentializing gendered experiences or asserting that women 'know' differently as a result of their biology. As Anderson argues, feminist epistemology aims to demonstrate and "explain what it is for a scientific theory or practice to be sexist and androcentric, how these features are expressed in theoretical inquiry and in the application of theoretical knowledge, and what bearing these features have on evaluating research." It also intends to defend "feminist scientific practices, which incorporate a commitment to the liberation of women and the social and political equality of all persons" (Anderson 1995a, b, 51).

As such, science should be understood as a socio-political endeavour supported by empirical scientific practice and guided by norm and values. Geoengineering science is no exception. It too is subject to assumptions and values that shape the course of its evolution. As it stands, these values, which include various levels of hierarchy, mono-causality, globalism, independence, and asociality, do not measure up to or satisfy Longino's complex framework. Again, it is not the case that values detract from or undermine valid science – rather, they are constitutive of science itself. Feminist science provides the resources for established theories to be questioned, power dynamics to be challenged, and traditional methodologies to be contested as well as forcing epistemology to take into account ideas, phenomenon and peoples that have been neglected and wilfully ignored.

Longino's virtues, taken together, work to provide a pathway by which science can satisfy the requirements of an empirical nature based on reason and the scientific method, yet continue to be grounded in context, difference, and diversity. This type of empiricism embraces "a form of contextualism that understands knowledge as the historical product of interactions between contextual factors such as social needs, values and traditions, and practices of inquiry such as observation, experiment, and reasoning" (Longino 1990, 176–177). These are validated by a community and satisfies local standards. It is the community of inquirers that guard against subjectivism and relativism while preserving the social nature of knowledge.

Empirical adequacy, ontological heterogeneity, mutuality of interaction, novelty, diffusion of power, and applicability to human needs are the principal, socially generated virtues that serve as guideposts for evaluating and guiding scientific practice. My objective in this book has been twofold: first, to assess, analyze, and challenge the assumption, values, norms and practices that underlie climate engineering using feminist empiricist approach; and second, in doing so, to formulate a novel critique of geoengineering that contributes to the existing literature that question its value as a means by which to mitigate climate disruption.

The variegated subjects touched upon in the course of assessing sulphate geoengineering – particularly with respect to its detachment from Longino's virtues – include the subjects of risk; scientific certainty; boundary objects; causality; marginalization; epistemological binaries (local/global, urban/rural, producer/consumer); modeling, idealization, and visualization; innovative methodologies; pluralism; expert versus lay knowledge; power dynamics and activism; commercialization; social justice; ethics; and intergenerationality. Each of these themes have been taken up in relation to climate engineering and used to support the argument that FCE's framework for desirable scientific practice, as constituted by Longino's scientific virtues, have not been met. In doing so, I have provided a uniquely feminist set of arguments against geoengineering not presently discussed in current academic discourse. The contributions of standpoint theory and techno-feminism further this analysis.

Overall, because climate engineering does not reflect the criteria or general ethos with which FCE approaches scientific practice, including everything from chosen research questions, guiding assumptions, variable choice, normative goals, political objectives, and preferred models, the most obvious conclusion is that a more comprehensive reassessment of it is necessary. This must take place in labs and amongst scientists involved in geoengineering research in order to bring new information, knowledge and perspectives to the field – which should be the aim of science. Curiously, it turns out that the very systems and frameworks that democratize science by challenging its historical androcentrism and encouraging diversity and dissent also happen to be feminist practices. It is not "that feminist values produce better science" per se, but that "good criticism produces better science, and feminist values create a fruitful stance for criticism of contemporary science, which is, for historical reasons, mostly steered by non-feminists" (Solomon 2012, 439).

The purpose of this book is to excite and encourage a more variegated and diverse use of feminist science studies to examine the most recent technological advances. Geoengineering technologies in particular have the potential to irreversibly alter the natural world in ways not experienced before and, as such, necessitate the most diverse, novel and pioneering modes of analysis possible. Hopefully this work goes some way in that direction.

References

Anderson, E. (1995a). Feminist epistemology: An interpretation and a defense. *Hypatia, 10*(3), 50–84.

Anderson, E. (1995b). Knowledge, human interests, and objectivity in feminist epistemology. *Philosophical Topics, 23*(2), 27–58.

Bardzell, S., & Churchill, E. F. (2011). IwC special issue "Feminism and HCI: New perspectives" Special Issue Editors' introduction. *Interacting with Computers, 23*(5), iii–ixi.

Bee, B. A., et al. (2015). A feminist approach to climate change governance: Everyday and intimate politics. *Geography Compass, 9*(6), 339–350.

Brown, H. P. (2011). Gender, climate change and REDD+ in the Congo Basin forests of Central Africa. *International Forestry Review, 13*(2), 163–176.

Buckler, S. (2010). Normative theory. In D. Marsh & G. Stoker (Eds.), *Theory and methods in political science* (pp. 156–180). New York: Palgrave Macmillan.

Cockburn, C. (1992). The circuit of technology: Gender, identity and power. In E. Hirsch & R. Silverston (Eds.), *Consuming technologies: Media and information in domestic spaces* (pp. 33–42). New York: Routledge.

Ellul, J. (1964). *The technological society*. New York: Vintage.

Gaard, G. (2011). Ecofeminism revisited: Rejecting essentialism and re-placing species in a material feminist environmentalism. *Feminist Formations, 23*(2), 26–53.

Goreau, T., et al. (2014). *Geotherapy: Innovative methods of soil fertility restoration, carbon sequestration, and reversing CO_2 increase*. Boca Raton: CRC Press.

Haack, S. (1996). Science as social?-yes and no. In J. Nelson (Ed.), *Feminism, science, and the philosophy of science* (pp. 79–93). Dordrecht: Springer.

Haack, S. (2009). *Evidence and inquiry: A pragmatist reconstruction of epistemology*. Amherst/ New York: Prometheus Books.

Haraway, D. (1988). Situated knowledges: The science question in feminism and the privilege of partial perspective. *Feminist Studies, 14*(3), 575–599.

Haraway, D. (1990). A manifesto for cyborgs: Science, technology, and socialist feminism in the 1980s. In L. Nicholson (Ed.), *Feminism/postmodernism* (pp. 190–233). New York: Routledge pp.

Haraway, D. J. (1994). A game of cat's cradle: Science studies, feminist theory, cultural studies. *Configurations, 2*(1), 59–71.

Haraway, D. (2013). *Simians, cyborgs, and women: The reinvention of nature*. New York: Routledge.

Harding, S. (1986a). *The science question in feminism*. Ithaca: Cornell University Press.

Harding, S. (1986b). The instability of the analytical categories of feminist theory. *Signs: Journal of Women in Culture and Society, 11*(4), 645–664.

Harding, S. (1989). Feminist justificatory strategies. In *Women, knowledge and reality* (pp. 189–201). Boston: Unwin Hyman.

Harding, S. (1992). Rethinking standpoint epistemology: What is" strong objectivity?". *The Centennial Review, 36*(3), 437–470.

Harding, S. (1995). "Strong objectivity": A response to the new objectivity question. *Synthese, 104*(3), 331–349.

Harding, S. (2004a). A socially relevant philosophy of science? Resources from standpoint theory's controversiality. *Hypatia, 19*, 25–47.

Harding, S. G. (2004b). *The feminist standpoint theory reader: Intellectual and political controversies*. Hove: Psychology Press.

Harding, S. (2008). *Sciences from below: Feminisms, postcolonalities, and modernities*. Durham: Duke University Press.

Harding, S. (2015). *Objectivity and diversity: Another logic of scientific research*. Chicago: University of Chicago Press.

Hartsock, N. C. (1998). *The feminist standpoint revisited and other essays*. Boulder: Westview Press.

Hartsock, N. (2004). The feminist standpoint: Developing the ground for a specifically feminist historical materialism. In S. Harding (Ed.), *The feminist standpoint theory reader*. London/New York: Routledge.

Hawkesworth, M. (2006). Grappling with claims of truth. In M. Hawkesworth (Ed.), *Innovation, feminist inquiry: From political conviction to methodological innovation*. London: Rutgers University Press.

Hekman, S. (1999). Identity crises: Identity, identity politics, and beyond. *Critical Review of International Social and Political Philosophy, 2*(1), 3–26.

Hekman, S. J. (2013). *Gender and knowledge: Elements of a postmodern feminism*. New York: Wiley.

Intemann, K. (2010). 25 years of feminist empiricism and standpoint theory: Where are we now? *Hypatia, 25*(4), 778–796.

Kaijser, A., & Kronsell, A. (2014). Climate change through the lens of intersectionality. *Environmental Politics, 23*(3), 417–433. http://www.tandfonline.com/doi/full/10.1080/09644 016.2013.835203. Accessed 6 Mar 2017.

Kofinas, G. (2002). Community contributions to ecological monitoring: Knowledge co-production in the U.S.-Canada Arctic borderlands. In I. Krupnik & D. Jolly (Eds.), *The Earth is faster now: Indigenous observations of Arctic environmental change* (pp. 54–91). Fairbanks: Arctic Research Consortium of the United States.

Levisohn, J. A. (2001). Inclusion and objectivity: Helen Longino's feminist theory of scientific inquiry. *Philosophy of Education Archive*, 337–345.

Longino, H. E. (1990). *Science as social knowledge: Values and objectivity in scientific inquiry*. Princeton: Princeton University Press.

Lykke, N. (2009). Non-innocent intersections of feminism and environmentalism. Women. *Gender and Research, 18*(3–4), 36–44.

Moller, H. F., et al. (2004). Combining science and traditional ecological knowledge: Monitoring populations for co-management. *Ecology and Society, 9*, 3.

Neis, B., & Williams, S. (1997). The new right, gender and the fisheries crisis: Local and global dimensions. *Atlantis: Critical Studies in Gender, Culture & Social Justice, 21*, 2.

Neumann, R. P., & Hirsch, E. (2000). *Commercialisation of non-timber forest products: Review and analysis of research*. Bogor: CIFOR.

Pinnick, C., et al. (Eds.). (2003). *Scrutinizing feminist epistemology: An examination of gender in science*. New Brunswick: Rutgers University Press.

Postman, N. (1993). *Technopoly: The surrender of culture to technology*. New York: Vintage.

Seager, J. (2009). Death by degrees: Taking a feminist hard look at the 2 climate policy. *Kvinder, Køn & Forskning, 3*(4), 11–21.

Solomon, M. (2012). The web of valief: An assessment of feminist radical empiricism. In Crasnow et al. (Eds.), *Out of the shadows: Analytic feminist contributions to traditional philosophy* (pp. 435–451). Oxford: Oxford University Press.

Tuana, N. (2013). Gendering climate knowledge for justice: Catalyzing a new research agenda. In M. Alston & K. Whittenbury (Eds.), *Research, action and policy: Addressing the gendered impacts of climate change* (pp. 17–31). Dordrecht: Springer.

Wajcman, J. (2000). Reflections on gender and technology studies: In what state is the art? *Social studies of science, 30*(3), 447–464.

Wajcman, J. (2002). Addressing technological change: The challenge to social theory. *Current Sociology, 50*(3), 347–363.

Wajcman, J. (2006). Technocapitalism meets technofeminism: Women and technology in a wireless world. *Labour & industry: A journal of the social and economic relations of work, 16*(3), 7–20.

Wajcman, J. (2007). From women and technology to gendered technoscience. *Information, Community and Society, 10*(3), 287–298.

Winner, L. (1977). *Autonomous technology: Technics-out-of-control as a theme in political thought.* Cambridge, MA: MIT Press.

References

163